Agriculture Issues and Policies

Agribusiness

Management, Opportunities and Challenges

AGRICULTURE ISSUES AND POLICIES

Additional books an ebooks in this series can be found
on Nova's website under the Series tab.

AGRICULTURE ISSUES AND POLICIES

AGRIBUSINESS

MANAGEMENT, OPPORTUNITIES AND CHALLENGES

RONALD LUNA
EDITOR

nova
science publishers
New York

Library of Congress Cataloging-in-Publication Data

ISBN: 978-1-53614-413-0

Published by Nova Science Publishers, Inc. † New York

Contents

Preface

The issue of how to assess the sustainability of governance structures is one of the most topical for researchers, farmers, investors, administrators, policymakers, interest groups, and the public at large. In the opening chapter of *Agribusiness: Management, Opportunities and Challenges*, the authors suggest a practical and holistic approach to assessing the sustainability of governing structures in the Bulgarian agri-business sector. Next, the authors propose that nanotechnology has the potential to significantly impact all sectors of agribusiness in the next decade, and may be used to enhance the possibility of re-developing conventional agribusiness resources. The objective of the closing chapter is to assess and measure the current growth on the agribusiness of coffee in Mexico, its trends and challenges, the technology being used, and the constraints and goals towards the sustainability of the value chain.

Chapter 1 - The issue of how to assess the sustainability of governance structures is one of the most topical for researchers, farmers, investors, administrators, policymakers, interest groups, and the public at large. In this chapter, the author suggests a practical and holistic approach to assessing the sustainability of governing structures in Bulgarian agri-business sector. First, justification of a new governance aspect of farm sustainability and on an approach for its integration into assessment is made. Following on from this, a specific framework, tailored to the contemporary conditions of Bulgarian agriculture, for assessing the sustainability of farms is proposed, including a system of principles, criteria, indicators, and reference values. After that, the overall governance, and economic, social, and environmental sustainability of major governance structures in Bulgarian agriculture are assessed and major factors for improving farming sustainability identified.

Chapter 2 - The rising demand for fast speed has generated the need for nanotech based systems and technologies, thus driving the growth of the nanotechnology markets. Nanotech finds application in different fields such as medical technology, production technology, lighting components and defense nanotechnics. Agribusiness sector is poised to be a 3 trillion-dollar industry in global investment by 2030 (World Bank 2013). Nanotechnology has the potential to impact significantly on all sectors of agribusiness ecosystem in the next decade. Nanotechnology could be used to enhance the possibilities of re-developing conventional and stranded agribusiness resources. Nanotechnology can make the industry considerably greener and competitive, with its current growth rate of 30% (US$ 1.08 billion) annually. Nanotechnology, holding a focus focusing on unique attributes of materials coming through from smaller sizes closer to nanometric, has the vibrancy to disrupt the agribusiness sector with an extension to other fields such as environmental engineering, biomedicine, defense and security, water utilization and conversion of energy. It is well noted that adoption of new technology and innovation is crucial in accumulation of global wealth and market value which now stand over US$ 1.2 trillion in a rough estimate. Nanotechnology has positioned itself as an advancement in technology that has the potential to create and revolutionize the whole agribusiness chain, with the ability to raise the bar for agricultural yield and efficiency, thus ensuring that there is sustainable food security and economic gain for key industrial players by 25% over and above the industry growth rate. It examines the strategies of adoption which companies use to obtain market shares. The structuring of the market under various market applications is also examined. Finally, the article looks at the positioning of the players with respect to economies of scale and competitive advantages.

Chapter 3 - Coffee is considered the economic engine for many of the producing countries, this way, coffee is the second most traded product in the world after oil, it is the most consumed beverage in the world, after water. The total surface of green coffee harvested in the world is more than 10 million hectares and 145 million bags, out of which 7% was originated in Mexico, which makes the country occupy the fifth place with 3.5% of the world coffee supply and thus has great productive roots and high consumption. Coffee-growing is carried out in 675 thousand hectares, with yields 382 kg ha-1 and which are concentrated in the areas of the Gulf Slopes, the Pacific Ocean Slopes, the Soconusco Region as well as the North Central Region of Chiapas and in other 12 states of the country, out of which 74% of the overall production is generated in Chiapas, Oaxaca and Veracruz, along with 404 municipalities, 4,572 localities in altitudes ranging from 300 to 2000 masl, but with indicated elevations of 600 to 1200 masl, and quite higher than other 510 producers. The production of coffee as agribusiness for the country is an important source of foreign exchange and income for thousands of small producers and around 30 indigenous groups, during the current harvest between the years of 2018 and 2019, there is an estimated production of 4.5 million kilos of coffee. Currently, a sustainable crop has been promoted that allows the use of coffee residues, creating added value as fertilizer, biofuel, substrate for various crops, and more recently, in the extraction of caffeine and polyphenols. The objective of this chapter is to assess and measure the current growth on the agribusiness of coffee in Mexico, its trends, its challenges, the technology being used, and of course, the constraints and goals towards the sustainability of the value chain and the utilization and application of the main coffee industry wastes (coffee pulp and husk) and the possibility of utilization of such, as raw materials for bio-refinery (energy, chemical and food) are discussed hereafter.

In: Agribusiness
Editor: Ronald Luna
ISBN: 978-1-53614-413-0

Chapter 1

A STUDY ON SUSTAINABILITY OF GOVERNANCE STRUCTURES IN THE BULGARIAN AGRIBUSINESS SECTOR

Hrabrin Bachev*
Institute of Agricultural Economics, Sofia, Bulgaria

ABSTRACT

The issue of how to assess the sustainability of governance structures is one of the most topical for researchers, farmers, investors, administrators, policymakers, interest groups, and the public at large. In this chapter, I suggest a practical and holistic approach to assessing the sustainability of governing structures in Bulgarian agri-business sector. First, justification of a new governance aspect of farm sustainability and on an approach for its integration into assessment is made. Following on from this, a specific framework, tailored to the contemporary conditions of Bulgarian agriculture, for assessing the sustainability of farms is proposed, including a system of principles, criteria, indicators, and reference values. After that, the overall governance, and economic, social, and environmental sustainability of major governance structures in Bulgarian agriculture are assessed and major factors for improving farming sustainability identified.

* Professor, Institute of Agricultural Economics, Sofia, Bulgaria, Email: hbachev@yahoo.com.

INTRODUCTION

The issue of how to assess the efficiency and sustainability of governing structures in agriculture is one of the most topical for researchers, farmers, investors, administrators, policymakers, interest groups, and the public at large around the globe (Andreoli and Tellarini, 2000; Bachev, 2005, 2006, 2016; Bachev and Petters, 2005; Bachev et al., 2016; Bastianoni et al., 2001; Candido et al., 2015; EC, 2001; FAO, 2013; Fuentes, 2004; Häni et al., 2006; OECD, 2001; Rigby et al., 2001; Sauvenier et al., 2005; UN, 2015). Despite enormous progress in the theory and practice of this new and evolving area of research, there is no consensus on how to define and assess the efficiency and sustainability of agrarian structures; what are the critical factors of sustainability of governance structures; and how to evaluate the level of sustainability of agrarian structures in a dynamic world where hardly anything is actually 'sustainable.'

In academic publications and official documents estimates of farm and agrarian sustainability are often unjustifiably equalized. Agrarian sustainability has a larger dimension than the sustainability of certain type of farming enterprise (family, cooperative, community farms etc.) (Bachev, 2010). For example, experiences from around the globe show that there are many highly sustainable farming enterprises that contribute little to overall agrarian sustainability - in Bulgaria there are numerous semi-market holdings and subsistence farms, large enterprise based on leased-in lands, and public farms etc. with low standards of environmental protection. On the other hand, sustainable agrarian development is commonly associated with the restructuring and adaptation of farming enterprises to constantly evolving markets, institutions, and natural environments. This process (pre)determines the low sustainability (unsustainability) and diminishing importance of farming enterprises of certain types (public, cooperative, small-scale), and the modernization of another part of them through diversification of their activity, the transformation of family farms into partnerships, agribusinesses, and vertically-integrated forms, etc.

Furthermore, in most cases a holistic approach is not applied, and the purely economic, purely productive, purely ecological, and purely social aspects of farm development are studied (assessed) independently from one another. Furthermore, the critical governance functions of farms, the costs associated with this governance (transaction costs), and the relations between different aspects of farm sustainability are mostly ignored (Bachev, 2004; Bachev and Petters, 2005). Nevertheless, very often the level of managerial (governance) efficiency and the adaptability of the farm predetermine its overall level of sustainability independently of its productivity, or its social or ecological responsibility/activity. Therefore, when assessing the sustainability and efficiency of different types of farm (subsistence, member-oriented, profit-making, part-time employment, conservation) one must also take into account their comparative potential in relation to the alternative market, private, and public (including informal) modes of governance of agrarian activity (Bachev, 2004).

Last but not least important, the majority of proposed frameworks for assessing sustainability apply a universal approach to 'faceless' farms, without taking into consideration the specificity of individual holdings (type, resource endowment, specialization, stage of development) and the environment in which they function (competition, institutional support and restrictions, environmental challenges and risks).

In this chapter, I suggest a practical and holistic approach to assessing the sustainability of governing structures in Bulgarian agri-business sector. First, a specific framework, tailored to the contemporary conditions of Bulgarian agriculture, for assessing the sustainability of governing structures is proposed, including a system of principles, criteria, indicators, and reference values. Particular emphasis is placed on the justification of a new governance aspect of farm sustainability and on an approach for its integration into assessment. After that, assessment of the overall, governance, economic, social, and environmental sustainability of farms of different juridical types in Bulgaria is made. Additionally, relevant factors of sustainable farming are identified.

THE FARM AS GOVERNANCE STRUCTURE

In the long term, no economic organization lasts if it is inefficient—it ends up being replaced by a more efficient form of organization (Bachev, 2004). Therefore, the problem of assessing the sustainability of farming enterprises is directly related to the estimation of factors and levels of farm efficiency (Bachev and Peeters, 2005). In traditional economics, the farm is presented as a production structure and analyses of efficiency are restricted to production costs (and the optimization of technological factors according to marginal rules). However, this approach fails to explain the high sustainability of diverse types of farms (subsistence, semi-market, cooperatives, small commercial, large agribusiness) with great variation in efficiency levels in Bulgaria, as well as around the globe (Bachev, 2004, 2010).

In addition to production costs, modern farming is also associated with significant transaction costs (Bachev, 2004). For instance, there are enormous costs in finding the best partners and prices; negotiating conditions of exchange and contract writing; enforcing and disputing agreements; and protecting property rights. As a rule, rational agrarian agents tend to seek, choose, and/or develop the most effective mode of organization (governance) for their relationships, minimizing their bounded rationality and safeguarding their investments and (absolute and contracted) rights from the hazard of opportunism. Consequently, in the long run, only effective governing structures that maximize benefits and minimize transaction costs tend to dominate (sustain) in agriculture (Bachev, 2004).

When transaction costs are high, they may block otherwise effective transactions, and restrict farm size far below a technologically optimal level. Very often, high costs for market trading (e.g., finding credit, marketing of output) and/or internal governance (deficiency of low transaction cost labor) limit a farming enterprise's size to miniature/subsistence farming, or family borders. In other instances, the existing effective potential to economize on market transaction costs causes a vast extension of farming enterprise size through backward, lateral, or forward integration of activity. For example, the high costs for market and contract trading under conditions of great economic and institutional uncertainty during the post-1989 transition in Bulgaria turned subsistence farming into the most effective (or only possible) form of organization available for the management of the agrarian assets (farmland, livestock, labor) of more than a million Bulgarians (Bachev and Tsuji, 2001). On the other hand, the enormous costs of market and standard contract trading have caused the dominance of integrated and interlinked modes and the concentration of commercial farming in a few thousands large firms and cooperative farms.

In a world of positive transaction costs, farms and other agrarian organizations have another significant economic role. They are not only productive, but also major governance structures—forms of organizing and governing transactions for the maximization of transaction benefits and the minimization of transaction costs. Therefore, the sustainability of individual types of farming enterprise cannot be correctly understood and estimated without analyzing their comparative production *and* governance potential.

The institutional framework is a critical factor for farm and agrarian sustainability (Bachev, 2004; Bachev and Peeters, 2005). The level of sustainability of a particular farming enterprise is quite different depending on the existence of public support programs; social mechanisms for protection from natural disasters and market failures; strict standards for product quality, working conditions, animal welfare, and environmental protection; permitted legal modes for market and private transactions; (in)efficiency of the system of laws and private contract enforcement; and diverse traditions. For instance, a strong dependency on public subsidies is often estimated to be an indicator of low sustainability for certain types of farming enterprise. At the same time, the experience of most developed countries with long-term subsidization of agriculture (EU, Switzerland, USA, Japan) indicates that high public support is an important factor in the strong sustainability of certain farming structures (e.g., family, smallholding, community, independent).

The operational goals of sustainable development and mechanisms for their achievement are also institutionally determined. For instance, (socially) acceptable formal and informal norms on the use of labor (employment of children, safety standards, minimum wages, social assurance); plants and livestock (animal welfare, preservation of biodiversity, use of GM crops); and environmental resources (water use rights, permissions for pollution of air, water, and soils), may all differ, even between the various regions of the same country. In Bulgaria, as in many other countries, there are different levels of compliance and enforcement of standards for food in market and subsistence holdings; different levels of social tolerance to (domestic) livestock farms in different residential locations; and significant differences in the implementation of environmental standards in different regions and communities.

The level of individual and overall transaction costs is also greatly determined by the institutional environment. Principally, if a state of law, trust, goodwill, and stability maintains in a society, then costs for the protection and exchange of private rights are insignificant (Bachev, 2004). Alternatively, if private rights are not well defined, enforced, or restricted, this limits the intensification of transactions and the optimization of farms. In Bulgaria, for a long time, the rights to certain resources have not been well defined (e.g., rights to comfort and a quiet environment, clean air, waters, roads, soil fertility and conservation) and this has created big difficulties in effectively allocating resources resulting in unresolvable disputes between polluting farmers on one side and neighbors, consumers, or visitors on another. Consequently, less sustainable structures have been used and some activities have not been carried out on a socially effective scale (Bachev, 2010).

As such, assessing the sustainability of farming enterprises must always take place in a specified and real institutional, rather than in an unrealistic normative or desirable, context (Bachev and Peeters, 2005). Nevertheless, the institutional aspect is commonly missing in most of the suggested systems for assessing agrarian and farm sustainability. Accordingly, non-feasible (ideal) norms, rather than real-life examples, are used as criteria for efficiency, such as: farming models from other (developed, neighboring) countries; assumptions like the existence of well-defined and enforced property rights and well-defined and enforced human and economic rights; the existence of an effectively working government and agrarian bureaucracy; a situation with no public intervention in market and private transactions. An assessment of sustainability must be based not on the above mentioned 'Nirvana' approach, but on an analysis of farming enterprises in their real institutional environment—the institutional framework (ecosystem) within which every farm has to be sustainable. Furthermore, any analysis must also incorporate the driving factors and trends in the evolution of the socioeconomic and institutional environment, such as the specific Bulgarian model for reforms and EU CAP implementation and the development of technologies, subsectors, and consumer preferences.

Economic exchanges allow for a more profitable use of resources, but also require additional (transaction) costs. Farmers and other economic agents (resource owners, consumers) tend to govern their relationships though the most effective forms—those which maximize their benefits and minimize their costs. Therefore, the most effective form and size of a farming enterprise is determined through optimization of the total (production and transaction) costs and trade-offs between the gain in productivity/benefits and the economy/gain in transaction costs (Bachev, 2004). As such, a farm will be efficient (sustainable) if it manages all transactions and activities in the most economic manner for the owner(s). Here, the focus is on the optimization of transaction costs, which ultimately determine the efficiency and sustainability of a farming enterprise. The optimization of production costs is a question of production management (agricultural techniques, crop rotation, adaptation to agro-climatic conditions), in which most farmers (farm managers) are competent or improve through (learning by doing) experience.

If a farm does not govern transactions (activity) effectively, it will become unsustainable, experiencing high costs and difficulties using institutions (as possibilities and restrictions) and carrying out transactions compared to other more feasible (alternative) forms of organization. In this case, there will be a strong incentive to explore the existing potential of the farming enterprise (adapting to a sustainable state) through a reduction or enlargement of its size, or via its reorganization or liquidation—either through an alternative farm or non-farm application of available resources; farm expansion through employment of additional resources; trade instead of internal use of owned land and labor; or through take-over or merger with another farm or business.

Transacting modes and acceptable net benefits will vary according to the particular preferences of individual agents, their entrepreneurship ability and experience, their risk aversion, and the opportunity costs of owned resources. Expected benefits may cover multiple dimensions: monetary or non-monetary income; profit; indirect revenue; pleasure of self-employment or family enterprise; enjoyment of agricultural activities; desire for involvement in the preservation of the environment, biodiversity, cultural heritage, or traditions; increased leisure and free-time, or other non-economic benefits (such as a desire to preserve the farm for heirs and future generations); or some combination of these.

In order to assess farm efficiency and sustainability, we have to put individual transactions at the center of our analysis and assess the level of associated costs and benefits according to the specific personal (behavioral) characteristics of participants (Williamson, 1996). As such, the analysis of sustainability is always specific to a particular farming enterprise. Major types of farm transactions are associated with: knowledge supply; innovation supply; land supply; labor supply; input supply; finance supply; insurance supply; and the marketing of services and products. Analysis needs to be made of the comparative organizational efficiency of every major transaction of a farm in its specific institutional, economic, and natural environment. If significant costs (difficulties) of some (any) type of transaction in relation to feasible alternatives are identified, then the farm should be considered unsustainable. Given the fact that an alternative form often diminishes one type while increasing other kinds of transaction costs and the widespread application of complex modes (e.g., interlinking credit supply with input supply and/or marketing of farm produce), the overall (internal and external) transaction costs of a farm have to be taken into account.

A farm's potential (internal incentives, real capability) for adapting to the evolving institutional, economic, and natural environment[1] through effective changes in governance (saving on transaction costs) and production structure (exploring technological possibilities for diversification and growth in productivity) needs to be estimated. If a farm does not have the potential to maintain or adapt to a more sustainable level(s), it will either be liquidated or transformed into another type of farm. For instance, if a farm faces enormous difficulties meeting institutional opportunities (access to public support programs) and restrictions (new quality and environmental standards, production quotas); or if it has serious problems supplying managerial capital (being a one-person farm with an aged farmer who has no successor capable or willing to continue the business); or if the supply of needed farmland (demand for non-agricultural use of land), or funding (insufficient finance, inability to sell equity or buy credit), or marketing output (changing demand for certain products, needs for cooperative members, strong competition with imported products) is limited, then it will not be sustainable, despite its historical or current efficiency. Therefore, the adaptability of a farm characterizes, to a great extent, its sustainability and this has to be used as the main criteria and indicator for its assessment.

Towards a More Precise Definition of Farm Sustainability

I believe that the sustainability of a farming enterprise has to be based on the literal meaning of that term and it should be considered the "ability to continue over time" (Hansen, 1996). In the long term, in order for a

---

[1] Changes in the natural environment are the slowest and affect to the smallest extent the short-term sustainability of an individual farm. On the other hand, the dynamics of economic conditions and institutional modernization represent a challenge for the majority of Bulgarian farms.

farming enterprise to be sustainable it has to be efficient, otherwise it will be replaced by a more efficient form of organization. As such, the analysis of farm sustainability has to include the evaluation of potentially effective different types of farming enterprise. What is more, analysis should not be restricted to technological efficiency (production costs and benefits), but incorporate efficiency of governance as well. The analysis of (structure and factors of) transaction costs gives us a key to understanding the efficiency and sustainability of typical Bulgarian agriculture forms—primitive holdings for subsistence and part-time employment; production cooperatives on the large scale; the organization of farming into huge businesses; and various mixed and 'grey' structures, in the period of post-communist transition since 1989.

It is necessary to evaluate the potential (internal incentives, real capability) of farming enterprises to adapt to evolving market, institutional, technological, and natural environments through effective changes in their governing forms (economy on transaction costs) and production structures (exploring technological possibilities for diversification and growth in productivity). If a farm is not adaptable to novel institutional opportunities (public support programs, free competition) and institutional restrictions (new rules, norms, standards for business, environmental protections) it will be unsustainable. Similarly, if a farm faces difficulties in organizing certain types of transactions, it will either be liquidated or transformed into another enterprise type. For instance, if a farm has high costs for supplying managerial capital, as in a one-person farm when an aged farmer has no successor, then it will not be sustainable in the future despite its historical or current efficiency.

As a production and management unit, the sustainability of a particular farming enterprise is determined by its activity and managerial decisions (efficiency, ability to adapt to evolving environments), and changes in the external environment (market dynamics and crisis, public support and restrictions, extreme climate). The latter are able to significantly improve or decrease the sustainability of individual enterprises, independent of the management decisions of a farm. For example, direct subsidies from the

EU have considerably increased the sustainability of many previously less sustainable Bulgarian farms (Bachev, 2010).

The high level of efficiency and sustainability of small-scale holdings for part-time employment and subsistence in Bulgaria and Eastern Europe cannot be properly evaluated outside analysis of the household and rural economy. Similarly, the high efficiency of cooperative farms during the post-communist transition period was not caused by their superior productivity when compared to family holdings, but on their capacity to organize activities with a high dependency (asset specificity) for members under conditions of great institutional and economic uncertainty (Bachev and Tsuji, 2001).

Finally, there exists no farm or any other system that is sustainable forever. Therefore, an assessment of the sustainability of a farming enterprise is also associated with the question: for how long? For what period of time are we talking about? Considering the constant evolution of the features and concept of sustainability and the evolution of the entire agrarian system, sustainability is increasingly perceived "as a process of understanding of changes and adaptation to these changes" (Raman, 2006). According to this new understanding, agrarian (and farm) sustainability is always specific to time, situation, and component, and characterizes the potential of agricultural systems to exist and evolve through adaptation to and incorporation of changes in time and space. For example, at the current stage of development respecting the rights of farm livestock and wild animals (animal welfare) is a substantial attribute of farm sustainability.

Moreover, the incorporated internal dynamism of the system also implies an 'end-life' and a particular agrarian system is considered to be sustainable if it achieves (realizes) its expected lifespan. For instance, if due to the augmentation of incomes of farm households the number of subsistence and part-time farms decreases while agrarian resources are effectively transferred to other novel or larger structures, this process should not be associated with a negative change in the sustainability of farming enterprises in a region or subsector. If a particular farming enterprise is not able to adapt to dynamic economic, institutional, and climatic changes through adequate modernization in technology, production, and organization, it is to be evaluated as having low sustainability.

The characterization of sustainability has to be system-oriented while the system has to be clearly specified in terms of its temporal and spatial boundaries, components, functions, goals, and hierarchical importance. This implies taking into account the diverse functions of farming enterprises at their contemporary stage of development, as well as their type, efficiency, and links (importance, dependency, and complementarity) to the sustainability (economy) of farming households, agrarian organizations, region, ecosystem, and sector (industry).

Sustainability has to reflect both the internal capability of a farm to function and adapt to its environment as well as the impact of constantly evolving socioeconomic and natural environments on the operation of an individual enterprise. However, the features of relatively independent (sub)systems need be clearly distinguished—e.g., while the satisfaction of engaging in farming activity is an important social attribute of a farming enterprise's sustainability, the modernization of social infrastructure and services in rural areas is merely a prerequisite (factor) for the long-term sustainability of individual farms.

Furthermore, an applied sustainability approach allows for a comparative analysis of diverse agricultural systems—farming enterprises of different types within a country; farms in different countries. As such, all approaches that associate comparability only with the continuous (quantitative) rather than the discrete properties of a system should be rejected (Hansen, 1996; Sauvenier et al., 2005). In fact, there is no reason to believe that the sustainability of an agricultural system can only increase or decrease. Discrete features (sustainable; non-sustainable) are possible, and of importance to farming enterprises managers, interests groups, and high-level decision and policy makers (Bachev and Peeters, 2005).

Any characterization of sustainability must also be predictive since it has to deal with future changes, rather than the past and the present. In addition, sustainability has to be a criterion for guiding changes in policy; in farming and consumption practices; in agents' behavior; and in research and development priorities. In that sense, the analysis of factors important in the historical sustainability of farms (the achieved level of sustainability) in a region, subsector, or country is extremely useful for the development of both theory and practice. Assessment of past states helps to define the approach, system, and important sustainability indicators, as well as identify critical factors and trends relating to the sustainability of farming enterprises. Efficient measures can be undertaken by managers, state authorities, and stakeholders to increase current and future sustainability levels through education, direct support, innovation, restructuring, and partnerships accordingly.

Last but not least, sustainability requires simple and rapid diagnosis and offers the possibility of intervention through identifying and prioritizing restrictions, testing hypotheses, and allowing for comprehensive assessment. This suggests that the concept of sustainability and its assessment needs to be easy to understand and practical to use without requiring huge costs. Here I raise the issue of economic justification for undertaking assessments or increasing their precision.

A system of adequate principles, criteria, and indicators for assessing the individual aspects and overall sustainability of farming enterprises according to the specific conditions of a particular country, subsector, region, or ecosystem needs to be worked out. Each element of such a hierarchical system needs to meet certain conditions (criteria) such as: discriminating power in time and space; analytical soundness; measurability; transparency; policy relevance; transferability to all types of farming enterprise; and relevance to sustainability. (Sauvenier et al., 2005). For instance, in Bulgaria, as in many other countries, when an assessment of a farming enterprise's sustainability is performed it is not considered important to include the greenhouse gas emissions of livestock and machinery. At the same time, the number of animals on a unit of farmland is of critical importance, since underutilization or over-exploitation of pasture, as well as the manner of storing and using manure, is critical to the sustainable exploitation of natural resources.

I believe that the definition of sustainability relevant to farming enterprises has to be based on the literal meaning of that term and perceived as a system of characteristics and the ability to continue over time. Sustainability has to be able to characterize all the major aspects of the activity of a farming enterprise—managerially sustainable, economically sustainable, socially sustainable, and ecologically sustainable (Figure 1-1).

Farming enterprise sustainability has to characterize the ability (internal potential, incentives, comparative advantages, importance, and efficiency) of a particular farm to maintain its standard of governance, and its economic, environmental, and social functions in the long term. A farming enterprise is sustainable if:

- It has good governance—that is to say it is the preferred form for the farmer/entrepreneurs (owners) and has the same or greater potential for ensuring the good governance of a farm's activities and transactions compared to other farms or economic organizations.
- It is economically viable and efficient—that is to say it allows for an acceptable economic return on the resources used and ensures the financial stability of the enterprise.
- It is socially responsible in relation to farmers, hired labor, other agents, communities, consumers, and society—that is to say it contributes towards the improvement of welfare and living standards of farmers and rural households, the preservation of agrarian resources and traditions, and the sustainable development of rural communities and society as a whole.
- It is environmentally friendly—that is to say its activity also aligns with the conservation, recovery, and improvement of the natural environment (land, water, biodiversity, atmosphere, climate, ecosystem) and the welfare of livestock and wildlife.

Depending on the combination of these four dimensions, the level of sustainability of a particular farming enterprise may be high, good, unsatisfactory, or it may be unsustainable. For instance, a farm may have a high level of governance and economic sustainability, and a low level of ecological and social sustainability. Nevertheless, a low or lack of sustainability of a farming enterprise in any of the four aspects (pre)determines the overall level of its sustainability—e.g., inferior governance equates to low overall sustainability.

The level of sustainability of a farming enterprise should be evaluated according to short-term (the program period), medium-term (the current generation of farmers), and long-term (the next generation) scales.

Assessment of the sustainability of farming enterprises has always to be made according to their specific socioeconomic and ecological context rather than an unrealistic (desirable, normative, ideal) context. In that sense, the employment of a 'Nirvana' approach in determining sustainability criteria (unrelated to the specific environmental scientific norms of agricultural techniques; farming models from other regions or countries; the assumptions of defined and enforced property rights and restrictions; an effectively working state administration; or a situation without missing markets and public interventions) is not correct. Taking into account external socioeconomic and natural factors allow us to identify the major factors that contribute to the sustainability of a particular farming enterprise: competitiveness; adaptability; development of farmers and agrarian organizations; access to public programs; level of state support; institutional environment; extreme climate; and plant and livestock diseases.

In the long term, there exists no economic organization that is not efficient—if it were, it would be replaced by a more efficient form of organization. Therefore, the problem of assessing the sustainability of a farming enterprise is directly related to the assessment of the level of governance, production, and ecological efficiency of farms. The potential of a farming enterprise to adapt to evolving markets, and the economic, institutional, and natural environment through effective changes in governing form, size, production structure, technology, and behavior has to be estimated. If the farm does not have the potential to sustain itself or adapt to a new and more sustainable level, its comparative efficiency and sustainability will decrease and eventually it will either be liquidated or transformed into another type of organization.

Framework for Assessing the Sustainability of Governing Structures in Bulgarian Agriculture

Farm sustainability is the ability (internal capability) of a particular farm to exist over time, maintain a good standard of governance, and maintain its economic, ecological and social functions according to the specific socioeconomic and natural environment in which it operates and evolves.

The sustainability of a farming enterprise has four aspects that are equally important and have always to be taken into account:

- *Managerial sustainability*—the farming enterprise has to have a good or high absolute and comparative standard of efficiency in the organization of its activity and (internal and external) relations. It should have a high level of adaptability to its evolving socioeconomic and natural environment according to the specific preferences (type of farm, character of production, long-term goals) and capabilities (training, experience, available resources, connections, power) of the owners of the farm.
- *Economic sustainability*—the farming enterprise has to have a good or high level of productivity in its utilization of natural, personal, material, and financial resources; an acceptable level of economic efficiency and competitiveness; and adequate financial stability.
- *Social sustainability*—the farming enterprise has to have good or a high level of social responsibility regarding farmers, workers, other agents, communities, and consumers. It should contribute to the conservation of agrarian resources and traditions; improve the welfare and living standards of farm households; and encourage the development of rural communities and society as a whole.
- *Ecological sustainability*—the farming enterprise has to have a good or high level of ecological responsibility and be associated with socially desirable conservation, recovery, and improvement of natural environments (landscape, land, water, biodiversity,

atmosphere, climate, ecosystem processes) and nature as a whole. It should respect animal welfare and other socially determined standards related to nature.

Assessing the sustainability of a farming enterprise can be done at different levels: an individual farm; farms of a particular type or kind; farms in a particular ecosystem; farms in a particular region; farms of a particular subsector of agriculture; all farms in a country; farms in different countries. Assessments at higher economic and special levels uses aggregates of individual farm assessments.

A system of selected (farm level or aggregated) indicators, which adequately reflect the major aspects of the sustainability of individual holdings, may also be used as a rapid diagnostic of farm sustainability at higher levels. For instance, the level of nitrate pollution in the groundwater of a region (ecosystem) can give good insight into the ecological sustainability of the farms in that region (ecosystem).

It is also necessary to estimate the importance of different kinds/types of farming enterprise in terms of their: overall resource utilization; total agricultural output; social and economic life; and their environmental impacts in relevant ecosystems, regions, subsectors, and agriculture as a whole. This determines the link between the sustainability of farms and agrarian sustainability, and makes it possible to take decisions on improving public policies and strategies for farms and agrarian organizations in terms of sustainable development

The level of sustainability of a farming enterprise and its contribution to overall agrarian sustainability usually depends on the farm's type. This requires classification of farming enterprises according to a number of criteria. The major types of farming enterprises according to their juridical status (formal registration) in Bulgaria, as specified by national legislation, are: physical (natural) person; sole trader; corporation; and cooperative. Furthermore, there are forms with open, close, mixed, or publicly traded membership. According to their type of ownership, farming enterprises may be private, state, municipal, community, public, local, foreign, or hybrid. According to the level of economic and managerial autonomy, there are independent, horizontally integrated, and vertically integrated holdings. According to market orientation, farms may be: subsistence holdings and farms for the servicing of members; semi-market farms; commercial farms; and business enterprises. An agricultural farm's size may be classified as: small-scale; medium-sized; and large. Different criteria may be used to classify them—the amount of land managed; the number of grazed livestock; the amount of employed labor; or gross income/economic size. According to production specialization, farms in Bulgaria may be classified in more or less aggregated groups: crop production (field crops, horticulture, permanent crops); livestock production (grazing livestock, pigs, poultry, and rabbits); or mixed production (mixed crops, mixed livestock, mixed crops and livestock). The ecological orientation and certification of a farm may include: organic certification or a transition period to organic certification; conventional production; ecological production; mixed production. The particular private or social objectives of a farm may include: experimental; demonstrative; educational; conservation and recovery of traditional breeds of livestock or varieties of crops; or protected and/or certified origins, products, services. According to location, farms may be classified into different groups depending on which ecosystem they include or are part of (plain, mountainous, semi-mountainous, riverside, seaside, protected zone and natural reserves, high risk), and/or which administrative (region, municipality, country), geographical zone (border, northern Bulgaria), or the social and economic (developed, developing, underdeveloped, unpopulated, declining) zones they are located in.

Assessing the sustainability of farming enterprises is done at a specific historical moment in time (a certain date). This inevitably reflects the existing knowledge and preferences of the farms and their impacts; the possibility of identifying, monitoring, measuring, and evaluating different aspects of their sustainability and impacts; the available information and access to first-hand farm data; and the needs of farm managers and agrarian policy at a particular moment of time. Two types of assessment of farm sustainability have to be distinguished: *Historical* (retrospective)—level and dynamics during a certain past-time period in the evolution of a farm. *Current* (actual)—giving an idea of the current state of a farm and its likely level of sustainability according to shorter or longer perspectives. Moreover, the *short-term*, *medium-term*, and *long-term sustainability* of farms should be distinguished and assessed.

The sustainability of a farm often changes over time, which necessitates estimation of its actual or likely level on a particular time-horizon: *Short-term*—the current program period of the implementation of EU CAP or 5 to 7 years. *Medium-term*—a longer period of times (e.g., 5 to 10 years); for a current assessment it is necessary to take into account the remaining time of the current generation of active farmers. The majority of Bulgarian farmers are of advanced age and are likely to retire in the coming decade—this is why it is appropriate to use a timeframe of 8 to 10 years for this kind of sustainability assessment. *Long-term*—a foreseeable longer-term of 10 to 15 or more years, which is also related to the conservation and transfer of farms and agrarian resources to subsequent generations.[2]

The hierarchical levels, which facilitate the formulation of a system for assessing the sustainability of farming enterprises, include well-determined and selected *principles*, *criteria*, *indicators*, and *reference values* (Figure 1).

- *Principles*—the highest hierarchical level associated with the multiple functions of agricultural enterprises. They are universal and represent the state of sustainability, which is to be analyzed according to the four main aspects—managerial, economic, social, and ecological.
- *Criteria*—these are more precise than the principles and are easily linked to sustainability indicators. They represent a resultant state of the evaluated farm when the relevant principle is realized.

---

[2] Assessment of farm sustainability in the very long term (25 to 30 or more years) is both difficult (impossible) and impractical since there is little (reliable) information about future trends, factors, preferences, impacts etc. For such long-term foresight, other methods of assessments are more appropriate (see COST, 2009); they are beyond of the scope of this study.

- *Indicators*—quantitative and qualitative variables of different types (behavior, activity, input, effect, impact), which can be assessed according to the specific conditions of the evaluated farms and allow the measurement of their compliance with particular criteria. The set of indicators provides a representative picture of farm sustainability in all its aspects. For instance, an indicator—'the extent of application of good agro-technics and crop rotation'—for the criterion—'soil erosion is minimized.'
- *Reference Values*—these are the desirable levels (absolute, relative, qualitative) for each indicator according to the specific conditions of the evaluated farming enterprises. They assist in the assessment of sustainability levels and give guidance for achieving (maintaining, improving) farm sustainability. They are determined by scientific, experimental, statistical, legislative, or other appropriate ways.

A reference value may be: *specific rule or standard (*e.g., application of good agricultural and ecological practices; labor safety standards; standards for animal welfare); *formal restriction (*e.g., norms for acceptable pollution of water, soil, and air; ecological limits for nitrate pollution of land and water); *norm for comparison* (e.g., optimum rate of chemical fertilization, pesticide application, water irrigation; extent of conservation of traditions); *minimum or maximum requirement (*e.g., lack of unsolvable problems for supplying necessary agricultural land and labor; optimum extent of a farming enterprise's liability); *limits of variation (*e.g., number of livestock on a unit of pasture; diversity of population of wildlife); *average values for similar enterprises (*e.g., average productivity and profitability of farms in a region or subsector); *trends (*e.g., level of income and welfare of rural households; greenhouse gas emissions from farms; insect and plant diversity); *personal or collective preferences (*e.g., satisfaction from farming; preservation of tradition; varieties and technologies).

Most of the reference values emphasize the level at which we can presume to guarantee the long-term sustainability of a farming enterprise.

Depending on the extent of their achievement, farming enterprises may have *high*, *good*, or *low sustainability*, or be *unsustainable*. For instance, farming enterprises with a higher than sectoral average profitability or lower soil acidity are more sustainable then others; farms with inferior profitability or greater soil acidity have lower economic or ecological sustainability, or are economically or ecologically unsustainable.

The reference values also characterize the condition of sustainability, deviation from which indicates a state of insufficient sustainability or unsustainability. For instance, farming enterprises that do not comply with official standards for labor (working, health and safety), animal welfare, the application of banned chemicals and technology, or the production of forbidden products (cannabis).

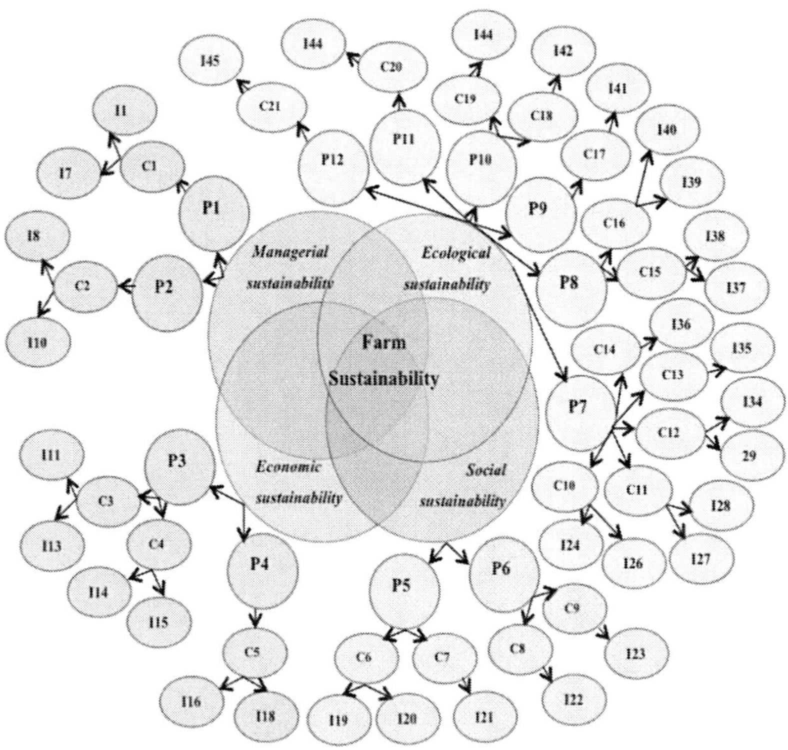

Source: Author.

Figure 2. System for Assessing the Sustainability of Farming Enterprises in Bulgaria.

Table 1. Principles, Criteria, Indicators, and Reference Values for Assessing the Sustainability of Farming Enterprises in Bulgaria

Principles	Criteria	Indicators	Reference Values
Governance Aspect			
Acceptable governance efficiency	Efficiency in governing an activity in relation to other feasible organizational forms	Comparative efficiency for supply and management of workforce	Similar to alternative organization
		Comparative efficiency for supply and management of natural resources	Similar to alternative organization
		Comparative efficiency for supply and management of short-term inputs	Similar to alternative organization
		Comparative efficiency for supply and management of long-term inputs	Similar to alternative organization
		Comparative efficiency for supply and management of innovations	Similar to alternative organization
		Comparative efficiency for marketing of products	Similar to alternative organization
		Comparative efficiency for supply and management of finance	Similar to alternative organization
Sufficient adaptability	Farm adaptability	Level of adaptability to market environment	Good
		Level of adaptability to institutional environment	Good
		Level of adaptability to natural environment	Good
Economic aspect			
High economic efficiency	Economic efficiency of resource utilization	Level of labor productivity	Similar to the average for the sector
		Land productivity	Similar to the average for the sector
		Livestock productivity	Similar to the average for the sector
	Economic efficiency of activity	Profitability of production	Similar to the average for the sector
		Farm Income	Acceptable by the owner
Good financial stability	Financial capability	Return on own capital	Average for the sector
		Overall Liquidity	Average for the sector
		Financial autonomy	Average for the sector
		Indicators	Reference Values

Principles	Criteria	Indicators	Reference Values
Social aspect			
Good social efficiency for farmer and farm households	Farmers welfare	Income per member of farm household	Similar to other sectors in the region
		Satisfaction of activity	Acceptable for the farmer
	Working conditions	Compliance with formal requirements for working conditions	Standards for working conditions in the sector
Acceptable social efficiency for not farmers	Preservation of rural communities	The extent farm contributes to preservation of rural communities	Overall actual contribution
	Preservation of traditions	The extent farm contributes to preservation of traditions	Overall actual contribution
Ecological aspect			
Protection of agricultural lands	Chemical quality of soils	Soil organic content	Similar to what is typical for the region
		Soil acidity	Similar to the average for the region
		Soil salinification	Similar to the average for the region
	Soil erosion	Extent of wind erosion	Similar to what is typical for the region
		Extent of water erosion	Similar to what is typical for the region
	Agro-technique	Crop rotation	Scientifically recommended for the region
		Number of livestock per ha.	Within acceptable numerical limits
		Rate of N fertilization	Within acceptable numerical limits
		Rate of K fertilization	Within limits of acceptable amounts
		Rate of P fertilization	Within limits of acceptable amounts
		Extent of application of good agricultural practices	Approved rules
	Waste management	Manure storage type	Rules for manure storage
	Water irrigation	Irrigation rate	Scientifically recommended rate for the region

Table 1. (Continued)

Principles	Criteria	Indicators	Reference Values
Protection of water	Quality of surface water	Nitrate content in surface water	Similar to what is average for the region
		Pesticide content in surface water	Similar to what is average for the region
	Quality of ground waters	Nitrate content in ground water	Similar to what is average for the region
		Pesticide content in ground water	Similar to what is average for the region
Protection of air	Air quality	Extent of air pollution	Acceptance from rural community
Protection of biodiversity	Variety of cultural species	Number of culturally important species	Similar to what is average for the region
	Variety of wild species	Number of wild species	Similar to what is average for the region
Animal welfare	Norms for animal welfare	Extent of compliance with animal welfare norms	Standards for animal breeding
Preservation of ecosystem services	Quality of ecosystem service	Extent of preservation of ecosystem services	Acceptance from communities

Source: Author.

These principles, criteria, indicators, and reference values have been formulated/selected by leading experts on farming enterprise sustainability. Moreover, they have to be permanently updated according to the specific conditions of evaluated farms and according to developments in science; measurement and monitoring methods; available information; industry standards; and social norms.

In this study, I have fully examined the available academic research, official documents, and experience in Bulgaria and other countries, and have carried out numerous consultations with leading national and international experts in the area. On this basis, a system that includes principles, criteria, indicators, and reference values relevant to contemporary conditions in Bulgaria has been formulated. An expert panel was set up with ten leading scholars working on the sustainability of farming enterprises. They discussed and evaluated the importance of the proposed principles, criteria, indicators, and reference values. Those considered most appropriate to the contemporary conditions of Bulgarian farming have been selected: 12 principles, 21 criteria, and 44 indicators with relevant reference values (Figure 2, Table 1).

A number of criteria were used in selecting indicators for the assessment of sustainability: relevance to reflecting aspects of sustainability; discriminatory power in time and space; analytical soundness; intelligibility and synonymy; measurability, governance and policy relevance; and practical applicability. For validation of these sustainability indicators, the widely used method of multi-criteria expert assessment was used, which is well presented in professional publications (Sauvenier et al., 2005). The goal was to select a balanced system with sufficient (around 10 for each aspect), but not too many indicators (no more than 50), ensuring efficiency of use.

To assess the sustainability level of individual farming enterprises it is necessary to use first-hand information provided by enterprise managers (for behavior, activities, results, objectives), available reports, statistical information, and expert assessments by professionals in the field. Usually there is a 'state of sustainability' of a farming enterprise with different values of a particular indicator and the level of sustainability needs to be specified. The experts have determined different qualitative states of sustainability (high, good, low, insufficient) and the state of unsustainability for diverse deviations from indicator values and reference values.

In order to visually present the diverse aspects and dimensions of sustainability of a particular farming enterprise, and integrate different types of indicators for a particular criterion, principle, and aspect of sustainability for one or more farms, the qualitative levels of each indicator are transformed into a unitless index of sustainability (IS_i). Very often individual indicators for each criterion and/or different criteria, principles, and aspects of sustainability are unequal. This significantly hardens the overall assessment and requires the integration of measurements (indicators, criteria, principles). The integral index for a particular criterion (SI(c)), principle (SI(p)), and aspect of sustainability (SI(a)), and the integral sustainability index (SI(o)) are arithmetic averages of the indices of composite indicators, criteria and principles, calculated by the following formulas:

$SI(c) = \sum SI(i)/n$ n - number of indicators in a particular criterion;
$SI(p) = \sum SI(c)/n$ n - number of criteria in a particular principle;
$SI(a) = \sum SI(p)/n$ n - number of principles in a particular aspect,
$SI(o) = \sum SI(a)/4$

It is well known that every integration of indicators of different types involves a lot of provisionality—this process implies equal importance and a certain interchangeability of the individual dimensions of sustainability. In particular, it presumes, that a low level of sustainability or a state of unsustainability for one (or several) indicator(s) could be compensated for by a higher value of another (or other) indicator(s) without changing the integral level of sustainability. However, this is not always true for the majority of indicators for managerial and economic sustainability in the short term, nor in the longer term for many of the indicators for social and ecological sustainability. For instance, a lack of good governance or economic sustainability rapidly makes the entire farming enterprise unsustainable independent of the level of social and environmental indicators.

According to the panel of experts, it is not necessary to give different weights to individual indicators when calculating the integral index for a particular criterion, principle, or aspect, or the overall level of sustainability. However, when the level of sustainability for any of the indicators is unsatisfactory or zero, its importance for the evaluated farm(s) should be analyzed. Furthermore, over longer periods of analysis the lowest level of sustainability for any indicator (criterion) will also (pre)determine the integral level of the particular aspect and the overall level of sustainability of the farming enterprise (Bachev, 2010).

The overall and particular (in terms of aspect, principle, criterion, indicator) sustainability of farming enterprises of a specific type and location is an arithmetic average of the figures for individual farms.

The integration of indicators does not diminish the analytical power of the model, but makes it possible to compare the sustainability of diverse aspects of an individual farming enterprise, of farms of different types, and of an entire sector. Since the assessment of sustainability levels for individual indicators is a (pre)condition of integration, the primary information is always available and may be analyzed in as much detail as is necessary. Depending on the objectives and final users of the analysis, the extent of integration of indicators needs to be differentiated. While farm enterprise managers, investors, and researchers may prefer detailed information for each indicator, more aggregated data for all farms and main aspects of sustainability are needed for decision-making at the highest level.

Study Method and Surveyed Farms

A survey of managers of representative market-oriented farms of different juridical types, sizes, product specializations, and ecological and geographical locations was used to assess the sustainability of farming enterprises in Bulgaria. The survey was carried out in the summer of 2016 with the assistance of the National Agricultural Advisory Service, the National Union of Agricultural Cooperatives, the National Association of Grain Producers, the Association of Decorative Plant Producers, the Association for Breeding Bulgarian Milk Sheep, and others. Professional associations were asked to identify representative farming enterprises of different types and locations to help give an adequate picture of farm sustainability.

The sustainability level of individual farming enterprises was based on estimates made by enterprise managers for each indicator according to four qualitative levels: high/higher or better than average for the sector/region; similar/good; low/lower or worse than average for the sector/region; negative/unsatisfactory/unacceptable. This approach was the only feasible way of getting the necessary data, since the levels for most governance and social indicators are practically known only by farming enterprise managers (e.g., satisfaction of activity, acceptable income, available alternatives for supply of inputs). The pilot study proved that the majority of farming enterprise managers were quite familiar with the productive, economic, and financial indicators of their farms in relation to the industry average, as well as with the state of most available environmental indicators. Precise data for certain ecological indicators on Bulgarian farms are not available as some can only be gathered through costly laboratory tests.

The qualitative estimates for individual farming enterprises were then quantified and transformed into sustainability indices for each indicator (SI(i)) using the following scales: 1 for high; 0,66 for good or average; 0,33 for low; and 0 for unsatisfactory or unacceptable. Equal weighting was given to each principle of a particular aspect, to each criterion of a particular principle, and to each indicator of a particular criterion to assess the sustainability of a farming enterprise for each criterion, principle, aspect, and overall. Four levels of sustainability of farming enterprises were distinguished by the panel of experts: high had a range of 0,84 to 1; good had a range of 0,5 to 0,83; low had a range of 0,22 to 0,49; and unsustainable had a range of 0 to 0,21.

The national official typology for farm holdings was used to classify the surveyed farming enterprises according to: juridical type; production specialization; geographical and administrative region; and ecological location. In addition, every surveyed manager determined his/her farming enterprise as: predominately for subsistence; small, medium-sized, or large for the sector; and located mainly in a plain, plain-mountainous, or mountainous region. This approach guarantees an adequate assessment as the farming enterprise managers are well aware of the specific and comparative characteristics of their holdings compared to others in the region and (sub)sector.

The survey of farming enterprises managers included 190 registered agricultural producers, comprising roughly 0,2% of all those registered under the 1999 Regulation No. 3 for Creating and Maintaining a Registry of Agricultural Producers in Bulgaria (MAF). Managers of representative farming enterprises of all juridical types, sizes, specializations, and location were surveyed (table 2). The structure and importance of the surveyed farming enterprises corresponds approximately to the real structure of registered agricultural producers and market-oriented holdings in Bulgaria.

Table 2. Type and Number of Surveyed Farming Enterprises (percent, number*)

Type and location	Physical Persons	Sole Traders	Cooperatives	Companies	Total
Total	80,00	4,21	6,84	8,95	190*
Mainly subsistence	11,18	0,00	0,00	0,00	8,95
Small size	57,89	37,50	0,00	5,88	48,42
Medium size	28,95	37,50	92,31	70,59	37,37
Big size	1,32	25,00	7,69	23,53	4,74
Field crops	10,53	25,00	69,23	29,41	16,84
Vegetables, flowers, and mushrooms	13,82	12,50	0,00	0,00	11,58
Permanent crops	24,34	25,00	0,00	11,76	21,58
Grazing livestock	17,76	25,00	0,00	5,88	15,79
Pigs, poultry, and rabbits	0,66	0,00	7,69	0,00	1,05
Mixed crops and livestock	14,47	0,00	23,08	23,53	15,26
Mixed crops	13,82	12,50	0,00	29,41	14,21
Mixed livestock	4,61	0,00	0,00	0,00	3,68
Mainly plain region	51,97	50,00	53,85	64,71	53,68
Plain-mountainous	19,74	50,00	38,46	17,65	22,11
Mainly mountainous	14,47	0,00	7,69	17,65	13,68
Lands in protected zones and territories	6,58	0,00	0,00	17,65	6,84
Mountainous regions with natural handicaps	15,13	0,00	7,69	11,76	13,68
Non-mountainous regions with natural handicaps	1,97	0,00	7,69	0,00	2,11
Northwest region	15,79	37,50	7,69	11,76	15,79
North-central region	21,05	0,00	23,08	23,53	20,53
Northeast region	15,13	12,50	38,46	11,76	16,32
Southwest region	14,47	0,00	7,69	11,76	13,16
South-central region	19,74	12,50	15,38	29,41	20,00
Southeast region	13,82	37,50	7,69	11,76	14,21

Source: Farm manager survey, July 2016.

Most of the participants in the survey were physical (natural) person enterprises and holdings of small or medium-size (for their sector), which specialized in field crops and permanent crops and were located mainly in the plain regions of the country. The largest number of surveyed farming enterprises were in the north-central and south-central regions of the country. The majority of surveyed physical person enterprises determined themselves to be small and medium-sized for the sector—a reasonable proportion were predominately for subsistence and a tiny part were large for the sector. The majority of physical person enterprises specialized in permanent crops; grazing livestock; a mix of crops and livestock; a mix of crops; vegetables, flowers, and mushrooms; and field crops. A small portion farmed mixed livestock, or pigs, poultry, and rabbits. The physical person enterprises surveyed were mainly located in plain and plain-mountainous regions; a smaller proportion were located in mountainous regions, mountainous regions with natural handicaps, land in protected zones and territories, and in non-mountainous regions with natural handicaps. The greatest number of surveyed physical person enterprises were situated in the north-central and in the south-central regions.

Around 37% of the surveyed sole traders were also of small or medium size for the sector. A quarter of these enterprises specialized in fields crops, permanent crops, or grazing livestock. 12% specialized in vegetables, flowers, and mushrooms, or mixed crops. Half of the sole traders were located in mainly plain regions and the other half were located in plain-mountainous regions; none of the surveyed enterprises were found in plain-mountainous regions, or regions with natural handicaps or land in protected zones and territories. Most of the sole traders were located in the northwest and southeast regions; the rest were located in the northeast and southwest regions of the country.

Corporations dominated the companies category with partnerships accounting for 5,88% of the total. The majority of companies self-determined as medium-sized for the sector, almost a quarter were large enterprises, and less than 6% were small. Most of these enterprises specialized in field crops or mixed crops. Another significant proportion specialized in a mixture of crops and livestock; smaller fractions specialized in permanent crops and grazing livestock. Most of the companies were located in plain regions and the rest were equally distributed between plain-mountainous and mountainous regions. A good number of these enterprises have land in protected zones and territories or are in mountainous regions with natural handicaps. The greatest proportion of companies were located in the south-central and north-central regions and the rest were evenly distributed between the other administrative regions of the country.

Most of the cooperatives surveyed were medium-sized for the sector and the rest were large. Most of them specialized in field crops. Another large share specialized in crops and livestock and the rest specialized in pigs, poultry, and rabbits. More than a half of the cooperative enterprises were located in plain regions. A large number were located in plain-mountainous regions and a smaller number were found in mountainous regions and in mountainous and non-mountainous regions with natural handicaps. The largest number of cooperatives were located in the northeast and north-central regions. Another sizeable proportion were in the south-central region and the rest were evenly distributed between the other three administrative regions of the country.

This survey found that the majority of farming enterprises are located in regions with normal economic, social, and environmental issues. However, a significant proportion of holdings are located in regions with large or extreme economic, social, and environmental challenges. A third of the managers indicated that their farm was located in a region with few or no environmental problems.

A good proportion of the managers were not aware of the character of the problems in their region or were not able to assess the level of socioeconomic and environmental challenges facing farming enterprises in the region where they were located. This is an issue related to the competency of farm managers in assessing environmental problems and social and economic challenges. The owners/managers of three-quarters of the surveyed farming enterprises were male and around 60% were under 55 years of age. The gender and age structure of the managers/owners will remain the same for the majority of Bulgarian farms over the coming 10 to 15 or more years and contribute to the sustainability of holdings.

A sizable number of the surveyed farming enterprises had existed for a relatively short period of time—up to 5 years—and 30% had existed for less than two years (Figure 3-3). A majority of holdings had had a reasonably long period of operation, including around 29% with 11 or more years of effective operation.

A little more than a half of the surveyed farm manager/owners indicated that they had made efforts to improve farm sustainability for up to 5 years. Another significant proportion had long-term experience of improving farm sustainability, including 19% with 11 or more years.

The majority of farming enterprises know well or very well the principles of good governance and economic sustainability. At the same time, most holdings acknowledged that their knowledge of the principles of social and environmental sustainability was at most satisfactory or entirely absent. A sizeable proportion of surveyed farming enterprises increased their capacity for managing sustainability through hiring a consultant.

The greatest internal capability for the management of diverse aspects of sustainability were shown by cooperatives—a considerable fraction knew the principles of good governance, and economic, social and environmental sustainability well or very well. Cooperative enterprises do not use consultants to improve their competence regarding sustainability. Knowledge regarding sustainability principles was also high among sole traders and companies; for physical person enterprises it was lower. Consultants were used to enhance the understanding of economic and environmental sustainability by 12% of sole traders and 9% of physical person enterprises.

Competency in sustainability principles increases with the size of a farming enterprise and larger holdings tend to have better knowledge of good governance, and economic, social, and environmental sustainability. There is also a marked difference in competence according to enterprise specialization. Those specializing in the following forms of production displayed greater competence in governance: field crops; grazing livestock; pigs, poultry, and rabbits; and mixed crops and livestock. Enterprises focused on pigs, poultry and rabbits and mixed crops and livestock had the greatest awareness of economic sustainability. Those specializing in mixed livestock production had the highest level of competence in environmental sustainability. The numbers of holdings with a high level of competence in sustainability principles was greatest among those with land in protected zones and territories and those located in the southwest region of the country. As such, in the future more efforts need to be directed towards improving the competence of farming enterprises in understanding and applying the principles of agrarian sustainability through education, training, consultation, and exchanging positive experiences.

Competency in the principles of sustainability is important; however, it is not enough to ensure effective management. Incomplete knowledge and a variety of other economic, technological, agronomic, and behavioral reasons mean that farmers do not always properly apply the principles of sustainable agriculture. According to most of the farm managers, they apply the principles of good governance, and economic, social, and environmental sustainability strictly or well. Nonetheless, a significant fraction of holdings only apply the principles of social, economic, and environmental sustainability, and good governance to a satisfactory standard. What is more, a number of enterprises indicated that they do not respect such principles at all (as much as 6% for social sustainability), or respect them only if potentially punished by legal sanctions for failure to follow them (as much as 8% for environmental sustainability).

The principles of agrarian sustainability are generally integrated and applied in the overall management system of cooperatives and companies. Around 8% of cooperatives apply the principles of environmental sustainability only if there are sanctions. Sole traders and physical person enterprises generally apply the principles of social sustainability. A good number of physical person enterprises respect the principles of sustainable agriculture only if there are sanctions—9% of them for environmental sustainability, 5% for economic sustainability, and 5% for good governance and social sustainability. All these data demonstrate that sanctions by state and local authorities encourage business behavior that promotes environmental sustainability for certain types of farming enterprise, like cooperatives and physical person enterprises.

The application of the principles of sustainability increases as the size of holdings increase—as a rule larger farming enterprises have better governance, and better economic, social, and environmental sustainability. Respecting sustainability principles is most common for farming enterprises specializing in field crops, grazing livestock, mixed crops and livestock, and mixed crops. This is also the case for holdings with land in protected zones and territories, those in non-mountainous regions with natural handicaps, and those in the southwest region of the country. For all groups of farming enterprises, the share of those that respect the principles of agrarian sustainability well or strictly surpasses the proportion of these that know these principles well or very well. As such, it is questionable how effectively some holdings apply these principles if they have limited knowledge of them.

SUSTAINABILITY LEVEL OF GOVERNING STRUCTURES

A multiple indicator assessment of the sustainability level of the surveyed farming enterprises indicates that the index of overall sustainability for the holdings is 0,55—this represents a good level of sustainability among Bulgarian farms (figures 4-1 and 4-2). The highest levels are shown in the indices of environmental (0,61) and social (0,57) sustainability, while indices of governance (0,52) and economic (0,5) sustainability are significantly lower. Improvement in these two areas appears critical to maintaining a good level of sustainability among Bulgarian farming enterprises.

Analysis of individual indexes for the primary sustainability principles, criteria, and indicators allows us to identify components contributing to aspects of sustainability among Bulgarian farming enterprises. For instance, good governance and economic sustainability among Bulgarian farming enterprises are both relatively low—the index of governance efficiency is 0,49 and the index of financial stability is 0,47 (Figure 4). Similarly, it is clear that despite overall environmental sustainability being relatively high, the index of preservation of agricultural lands (0,52) and the index of preservation of biodiversity (0,56) are relatively low.

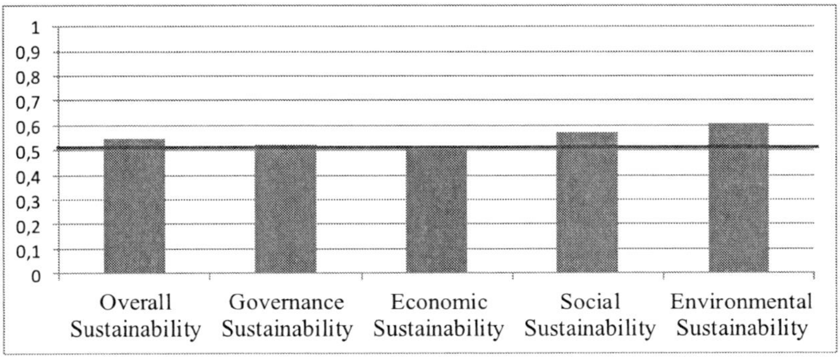

Source: Farm manager survey, July 2016.

Figure 3. Indices of Overall Sustainability, Good Governance, and Economic, Social, and Environmental Sustainability for Bulgarian Farming Enterprises.

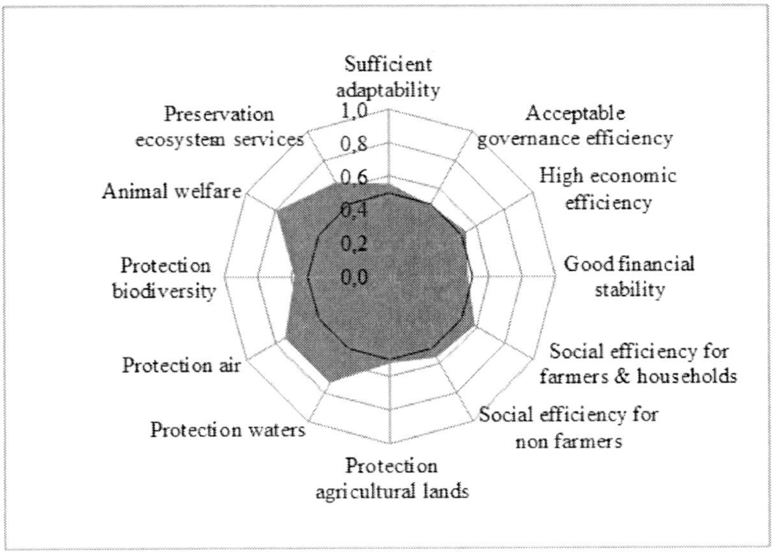

Source: Farm manager survey, July 2016.

Figure 4. Indices of Sustainability of Bulgarian Farming Enterprises for the Main Principles of Good Governance, and Economic, Social, and Environmental Sustainability.

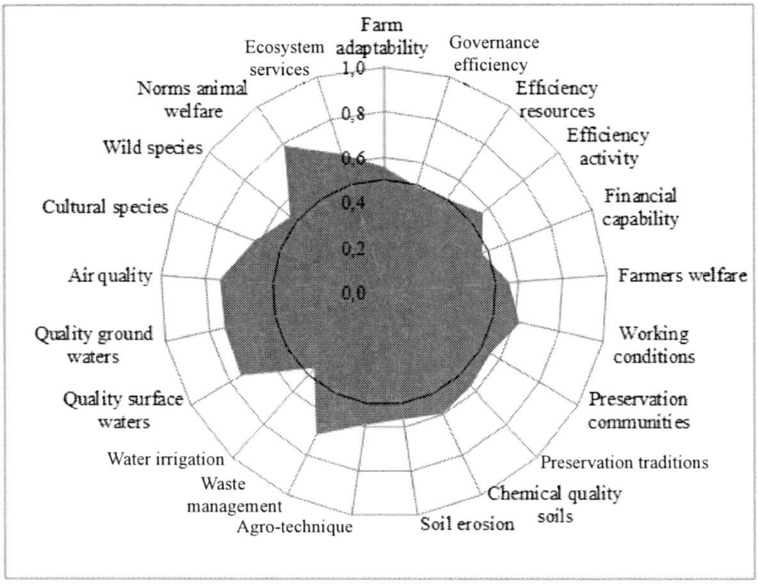

Source: Farm manager survey, July 2016.

Figure 5. Level of Sustainability of Bulgarian Farms for Individual Criteria of Governance, and Economic, Social, and Environmental Sustainability.

In depth analysis of individual criteria and indicators further specifies the elements that enhance or reduce a farming enterprise's sustainability. For instance, insufficient comparative governance efficiency and financial capability (Figure 5) is determined by a combination of low comparative efficiency in the supply of short-term inputs in relation to alternative organizations (0,28), unsatisfactory profitability of an enterprise's own capital (0,41), and low overall liquidity (0,48) (Figure 6).

Similarly, low values on the indices of preserving agricultural land and preserving biodiversity are determined accordingly by insufficient application of recommended irrigation norms (0,46), high level of soil erosion by water (0,55), and a reduced number of wild animals on farm territory (0,53).

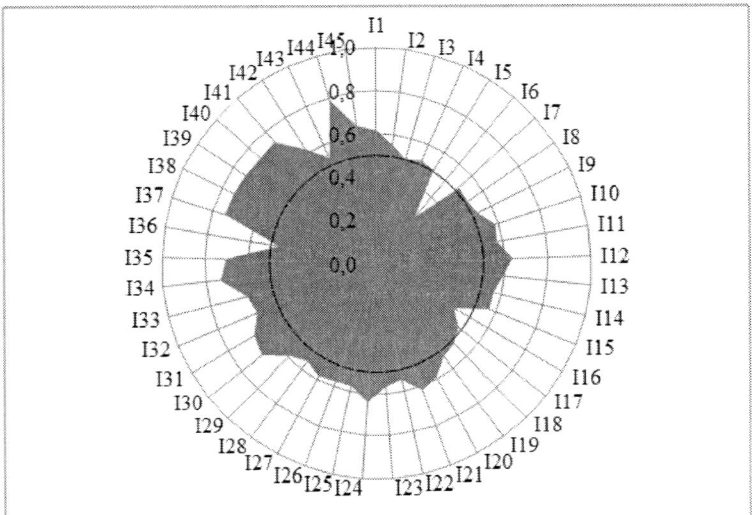

Source: Farm manager survey, July 2016.

Figure 6. Indicators* for Assessing the Sustainability of Bulgarian Farming Enterprises.

*I1-Level of Adaptability to Market Environment; I2-Level of Adaptability to Institutional Environment; I3-Level of Adaptability to Natural Environment; I4-Comparative Efficiency of Supply and Governance of Labor Resources; I5-Comparative Efficiency of Supply and Governance of Natural Resources; I6-Comparative Efficiency of Supply and Governance of Short-Term Inputs; I7-Comparative Efficiency of Supply and Governance of Long-Term Inputs; I8-Comparative Efficiency of Supply and Governance of Innovation; I9-Comparative Efficiency of Supply and Governance of Finance; I10-Comparative Efficiency of Governance of Marketing of Products and Services; I11-Land productivity; I12-Livestock Productivity; I13-Level of Labor productivity; I14-Rate of Profitability of Production; I15-Income of Enterprise; I16-Rate of Profitability of Own Capital; I17-Overall Liquidity; I18-Financial Autonomy; I19-Income per Farm-Household Member; I20-Satisfaction of Activity; I21-Compliance with Working Standards; I22-Contribution to Preservation of Rural Communities; I23-Contribution to Preservation of Traditions; I24-Nitrate Content in Surface Water; I25-Pesticide Content in Surface Water; I26-Nitrate Content in Ground Water; I27-Pesticide Content in Ground Water; I28-Extent of Air Pollution; I29-Number of Culturally Important Species; I30-Number of Wild Species; I31-Respect of Animal Welfare; I32-Preservation of Quality of Ecosystem Processes; I33-Soil Organic Content; I34-Soil Acidity; I35-Soil Salinification; I36-Extent of Wind Erosion; I37-Extent of Water Erosion; I38-Crop Rotation; I39-Number of Livestock per ha. of Farmland; I40-Nitrogen Fertilizer Norms; I41-Phosphorus Fertilizer Norms; I42-Potassium Fertilizer Norms; I43-Application of Good Agricultural Practices; I44-Type of Manure Storage; I45-Irrigation Rate

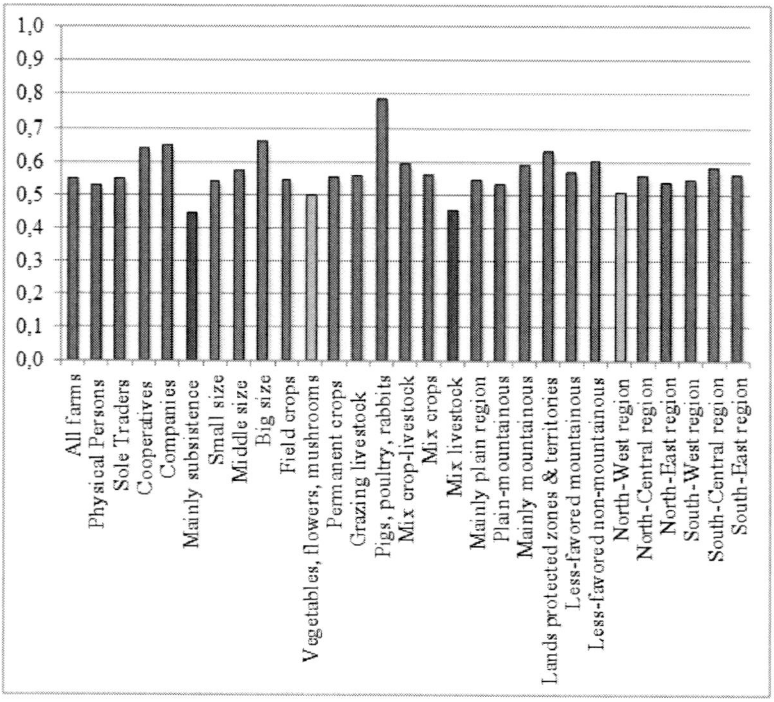

Source: Farm manager survey, July 2016.

Figure 7. Index of Sustainability of Bulgarian Farming Enterprises of Different Types and Locations.

Low values for indicators help identify specific areas that require improvement through changes in management strategy and/or public policy. For instance, despite the overall adaptability of farms being relatively high (0,56), the adaptability of farms to changes in the natural environment (climate, extreme events) is relatively low (0,5). Therefore, effective measures must be put in place to improve this through education, training, information, ameliorative agricultural techniques, restructuring of production, different crop varieties, and technological and organizational innovation.

The high levels of certain indicators show the absolute and comparative advantages of Bulgarian farms in terms of sustainable development. At the current stage of development, the most prominent of these include: respecting animal welfare standards; preserving surface and ground water quality through limiting contamination with nitrates and pesticides; preserving air quality; implementing good agricultural practices; reducing livestock per unit of farmland; ensuring acceptable labour conditions; comparative satisfaction from farming; optimal productivity of livestock; good adaptability to the market (prices, competition, demands); and comparative efficiency of governance of marketing of products and services.

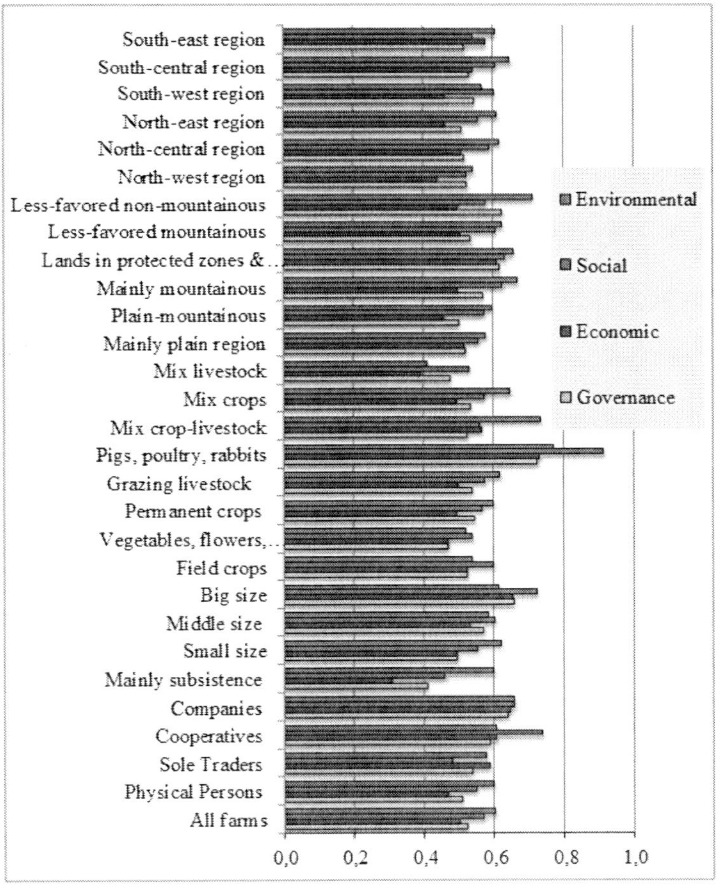

Source: Farm manager survey, July 2016.

Figure 8. Levels of Governance, and Economic, Social, and Environmental Sustainability of Bulgarian Farming Enterprises of Different Types and Locations.

There is significant variation in the level of sustainability of farming enterprises of different types and locations (Figure 7). Only those holdings predominately for subsistence and mixed livestock production have low sustainability. Good governance and economic and social sustainability are particularly low for subsistence farms (Figure 8). Mixed livestock farms tend to have low levels of governance and economic and environmental sustainability and only marginal social sustainability.

Holdings specializing in vegetables, flowers, and mushrooms had an overall sustainability of (0,5), with low levels of governance and economic sustainability (0,47). They had a relatively poor level of social sustainability (0,5) and environmental sustainability (0,52). Physical person enterprises had an overall sustainability level of (0,53), with low economic sustainability (0,47) and marginal social sustainability (0,5) and governance (0,51) (in contrast to their relatively good environmental sustainability). Farms located in the northwest region of the country had an overall sustainability of (0,5), with low economic sustainability (0,44) and not particularly good social sustainability (0,52), environmental sustainability (0,54), or governance (0,53). Effective measures need to be undertaken to improve all aspects of sustainability among these enterprises.

Small farming enterprises (0,496); those specializing in mixed crops (0,49) and permanent crops (0,498); and those holdings situated in plain-mountainous regions (0,46), mountainous regions (0,498), the northeast (0,46), and the southwest (0,46) all had low economic sustainability. Consequently, the overall sustainability of these farming enterprises is marginal. These data show that for all these enterprises effective measures need to be undertaken to increase their economic sustainability.

Sole traders had low social sustainability (0,48) and appropriate measures need to be introduced to improve this aspect through training, stimulation, regulation, and support. Companies, cooperatives, and large farming enterprises had high levels of overall sustainability with high levels of governance, and economic, social and environmental sustainability. Holdings specializing in the production of pigs, poultry, and rabbits had the highest level of sustainability, with very good levels of governance, and economic, social, and environmental sustainability. They are also the only type of enterprise to have a high level of social sustainability (0,9).

Farming enterprises with land in protected zones and territories and those located in non-mountainous regions with natural handicaps all had high levels of overall sustainability. Farming enterprises in non-mountainous regions with natural handicaps had reasonable levels of governance and environmental sustainability and low levels of economic and social sustainability. Those in the south-central region had relatively good levels of environmental and social sustainability, but poor governance and low economic sustainability. Farms with a mixture of crops and livestock had reasonable levels of environmental sustainability, but a lower level of governance.

SUSTAINABILITY INDICATORS FOR DIFFERENT TYPES OF FARMING ENTERPRISE

There was significant variation in the values for individual sustainability indicators for farming enterprises of different juridical types. Most sustainability indicators for physical person enterprises were low, highlighting their low sustainability for individual aspects and their overall lack of sustainability (Figure 9). These enterprises had poor levels of governance: level of adaptability to natural environment (0,49); comparative efficiency of supply and governance of labor resources (0,49); natural resources (0,49); long-term inputs (0,48); and innovation (0,49). They showed an extremely low comparative efficiency of supply and governance of short-term inputs (0,26). The economic sustainability of physical person enterprises is particularly low in respect of livestock productivity (0,34); rate of profitability of enterprise capital (0,36); overall liquidity (0,44); and financial autonomy (0,48). In terms of social sustainability, the only low component for these enterprises was income per farm-household member (0,49). In terms of environmental planning, low components included complying with norms for livestock per ha. (0,39); type of manure storage (0,39); animal welfare (0,43); and rate of irrigation (0,49). At the same time, a number of indicators of environmental sustainability were reasonably high for physical person enterprises: nitrate and pesticides in surface and ground waters; air pollution; and application of good agricultural practices. All these positive aspects of physical person enterprises need to be maintained and encouraged, while other indicators for ecological efficiency should be improved in order to ensure better overall sustainability among these holdings.

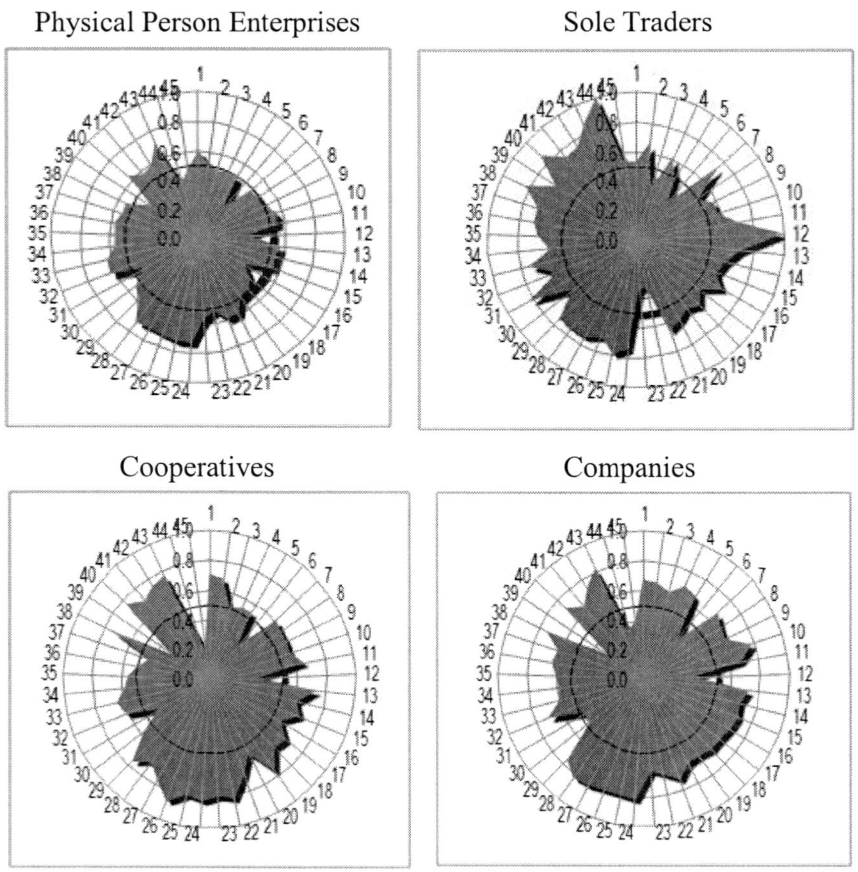

Source: Farm manager survey, July 2016.

Figure 9. Sustainability Indicators for Farming Enterprises of Different Juridical Types in Bulgaria.

Sole traders had low values for governance in terms of their level of adaptability to the natural environment (0,37) and comparative efficiency of supply and governance of short-term inputs (0,33). In terms of social sustainability, low values were seen for their contribution to the preservation of rural communities and preservation of traditions (0,33) (Figure 9). Sole traders, however, had high sustainability in ecological aspects in relation to type of manure storage; nitrogen fertilization norms; and application of good agricultural practices. Their implementation of effective crop rotation was lower. Enterprises with livestock showed a high level of sustainability for livestock productivity as well as a reasonable level in terms of their respect for animal welfare standards. Sole traders showed reasonably good values for many indicators of environmental sustainability: nitrate and pesticides in surface and ground waters; air pollution; number of culturally important species; organic content of soil; extent of wind and water erosion; and application of recommended amounts of potassium and phosphorus fertilizer. Sole traders also showed high values for: comparative efficiency of supply and governance of long-term inputs; level of labour productivity; and level of land productivity. All these aspects contribute to the development of good governance and economic sustainability.

Cooperatives showed good values for governance and social and economic sustainability: level of adaptability to market environment; level of labour productivity; income per farm-household member; contribution to the preservation of rural communities and traditions (Figure 9). Cooperative enterprises had high values for numerous environmental indicators, high ecological sustainability in terms of nitrates in ground water, and good ecological sustainability in terms of nitrates and pesticides in surface water; pesticides in ground water; number of culturally important species; application of good agricultural practices; efficient crop rotation; and application of nitrogen and phosphorus fertilizers. Cooperatives were environmentally unsustainable in terms of their use of irrigation (0,2) and had low values for: comparative efficiency of supply and governance of short-term inputs (0,3); livestock productivity (0,33); livestock per ha. (0,31); type of manure storage (0,31); animal welfare (0,41); and extent of water erosion (0,43). These areas of activity need to be improved considerably in order to increase the level of governance and sustainability of cooperatives.

Companies had the highest values in terms of standard of governance: comparative efficiency of supply and governance of labour resources and comparative efficiency of governance of marketing of products and services (Figure 9). In terms of economic sustainability, the highest values were seen for labor productivity and enterprise income. In terms of social sustainability, the highest value was for compliance with working standards. In terms of environmental suitability, high values were shown for: nitrates and pesticides in surface and ground waters; air pollution; application of good agricultural practices; efficient crop rotation; number of culturally important species; application of nitrogen and phosphorus fertilizers according to norms; and preservation of the ecosystem. Companies had low values for indicators of governance and economic sustainability: comparative efficiency of supply and governance of short-term inputs (0,35) and livestock productivity (0,35). Indicators for ecological sustainability with low values included: livestock per ha. (0,29); type of manure storage (0,35); animal welfare (0,41); rate of irrigation (0,41); and number of wild species on farm territory of (0,49). These areas of activity need to be improved in order to increase the standard of governance, and the economic, environmental, and overall sustainability of corporate enterprises.

Analysis of the Sustainability of Different Governance Structures

Physical person enterprises are the most numerous kind of farming enterprise (three quarters of all surveyed holdings) and to a large extent they (pre)determine the 'average' sustainability level for all enterprises in Bulgaria. Consequently, the level of overall sustainability of physical person enterprises of different types deviates insignificantly from the average sustainability levels of different categories of enterprise (Figure 10).

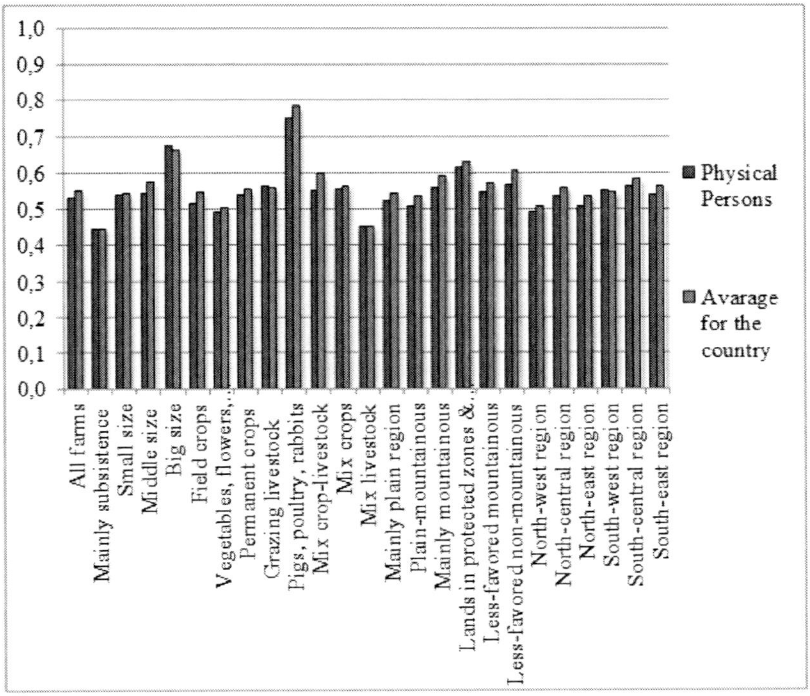

Source: Farm manager survey, July 2016.

Figure 10. Sustainability of Physical Person Enterprises of Different Types in Bulgaria.

There are significant variations in the sustainability of physical person enterprises according to their size, specialization, ecology, and geographical location. Product specialization and enterprise location appear to be more important factors for the sustainability of physical person enterprises than their juridical status. The greatest sustainability among physical person enterprises was shown by those of large size specializing in pigs, poultry, and rabbits, with land in protected zones and territories, and located in the south-central region of Bulgaria. At the same time, very low sustainability was shown among physical person enterprises that were predominantly for subsistence; those specializing in mixed livestock and in vegetables, flowers, and mushrooms; and those located in the northwest region. In terms of ecology, the lowest sustainability among physical person enterprises was found among those situated in plain-mountainous regions of the country.

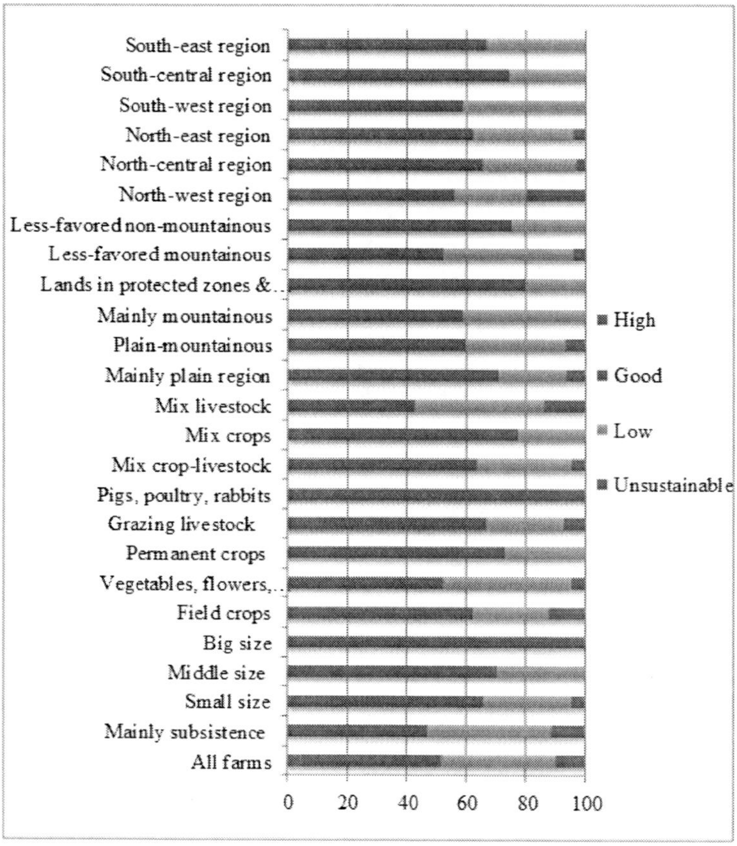

Source: farm manager survey, July 2016.

Figure 11. Structure of Physical Person Enterprises of Various Types with Different Levels of Sustainability in Bulgaria (percent).

There was also a significant difference in sustainability between the main types of physical person enterprise (Figure 11). All physical person enterprises of large size and specializing in pigs, poultry, and rabbits; and most of those specializing in mixed crops and permanent crops and located in non-mountainous regions with natural handicaps or with land in protected zones and territories, had good or high levels of sustainability.

Most physical person enterprises predominantly for subsistence and those with mixed livestock had low sustainability or are unsustainable. Low sustainability or unsustainability was shown by a considerable proportion of physical person enterprises specializing in: vegetables, flowers, and mushrooms; grazing livestock; and crops and livestock. This was also the case with those located in mountainous regions with natural handicaps, those in plain-mountainous regions, and those in the northwest and the southwest regions of the country.

Sole traders showed variation in sustainability depending on their size, specialization, ecology, and geographical location. Sole traders with the highest sustainability were those of large size specializing in vegetables, flowers, and mushrooms, and located in plain regions and the south-central region of Bulgaria (Figure 12). Sole traders with low sustainability included those specializing in mixed crops and grazing livestock; those of small size; those located in plain-mountainous regions; and those located in the northwest region of the country.

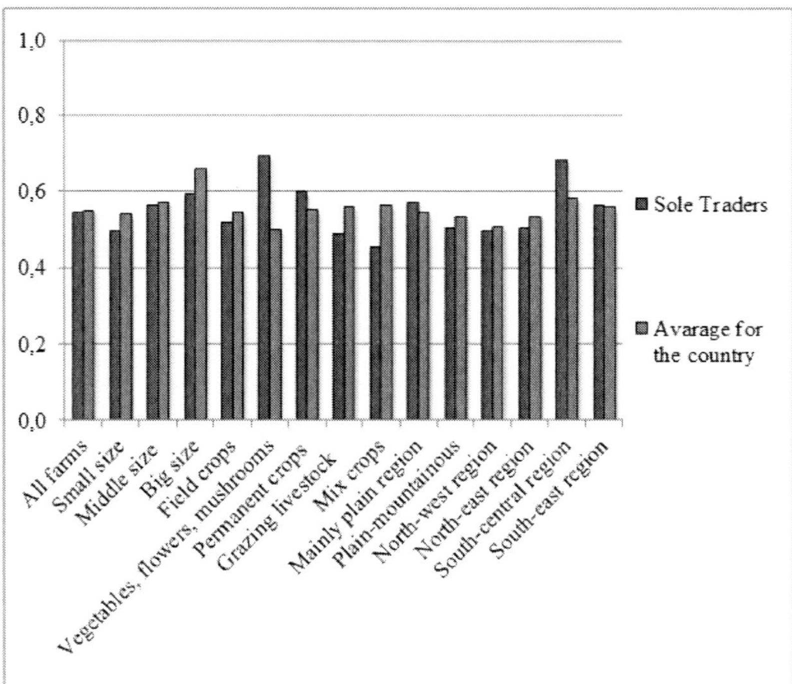

Source: Farm manager survey, July 2016.

Figure 12. Sustainability of Sole Traders of Different Types in Bulgaria.

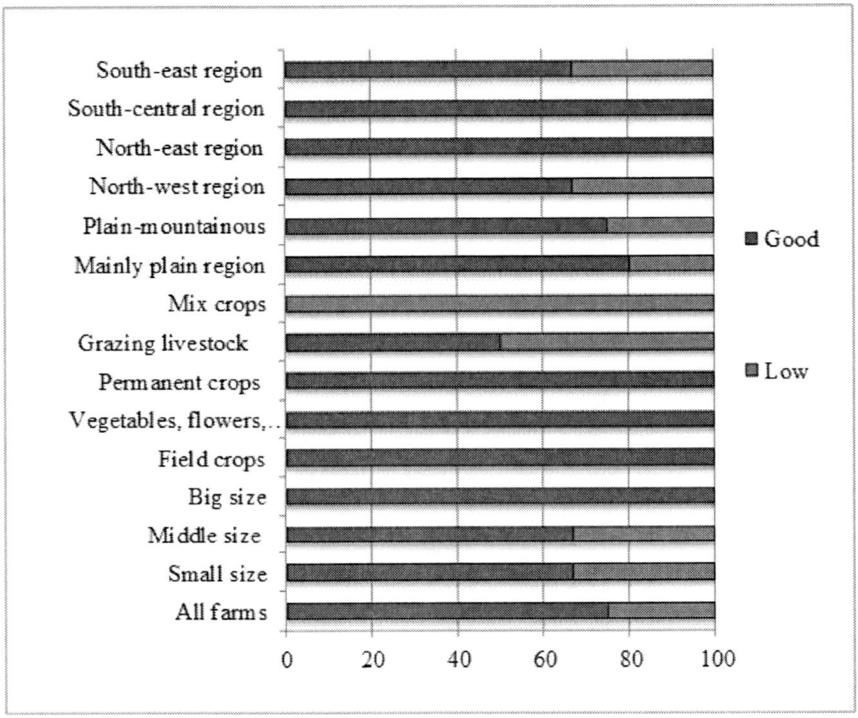

Source: Farm manager survey, July 2016.

Figure 13. Structure of Sole Traders of Various Types with Different Sustainability Levels in Bulgaria (percent).

Sole traders at the lowest and highest level of sustainability deviated significantly from the average levels of sustainability for the various categories of farming enterprise. This shows that the specific juridical status of the category of sole trader is a critical factor in determining the level of sustainability in this group, rather than the enterprise belonging to a certain type.

Among the remaining sole traders, levels of sustainability were close to the average for Bulgaria and this shows that for these sole traders size, specialization, and location are important factors in their sustainability.

There was significant variation in the share of sole traders of different types with good to low sustainability levels (Figure 13).

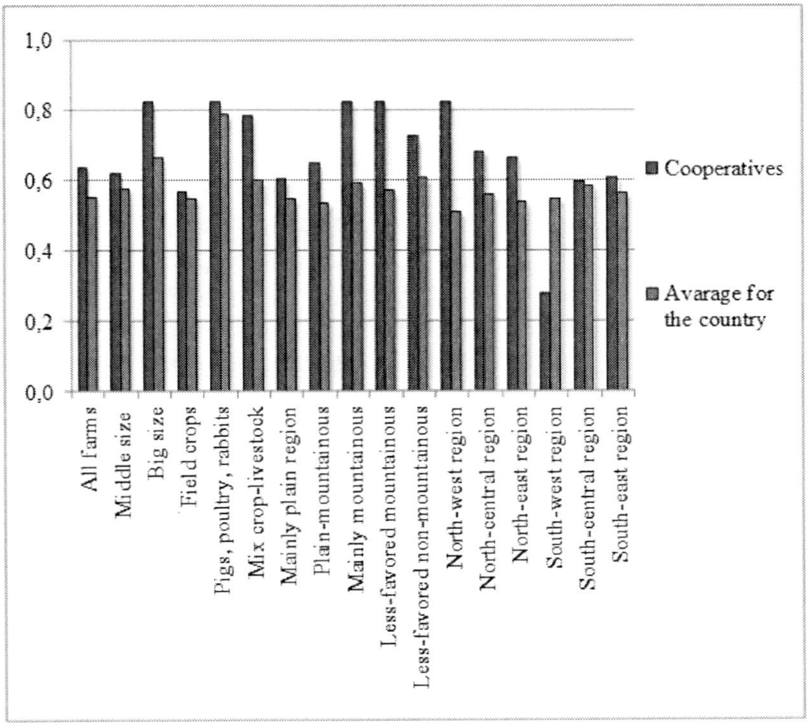

Source: Farm manager survey, July 2016.

Figure 14. Sustainability of Cooperatives of Different Types in Bulgaria.

All farming enterprises of large size; specializing in field crops, vegetables, flowers, and mushrooms, or permanent crops; and located in the northeast and south-central regions of the country had good sustainability. On the other hand, all holdings specializing in mixed crops; half of those specializing in grazing livestock; and one third of these of small and medium size, as well as those situated in the northwest and southeast regions of the country had low sustainability.

There exists considerable difference in sustainability depending on size, specialization, and location for cooperatives. Those with the best sustainability (borderline high) were cooperatives of large size; specializing in pigs, poultry, and rabbits; and located in mountainous regions, mountainous regions with handicaps, and the north-central region of the country (Figure 14).

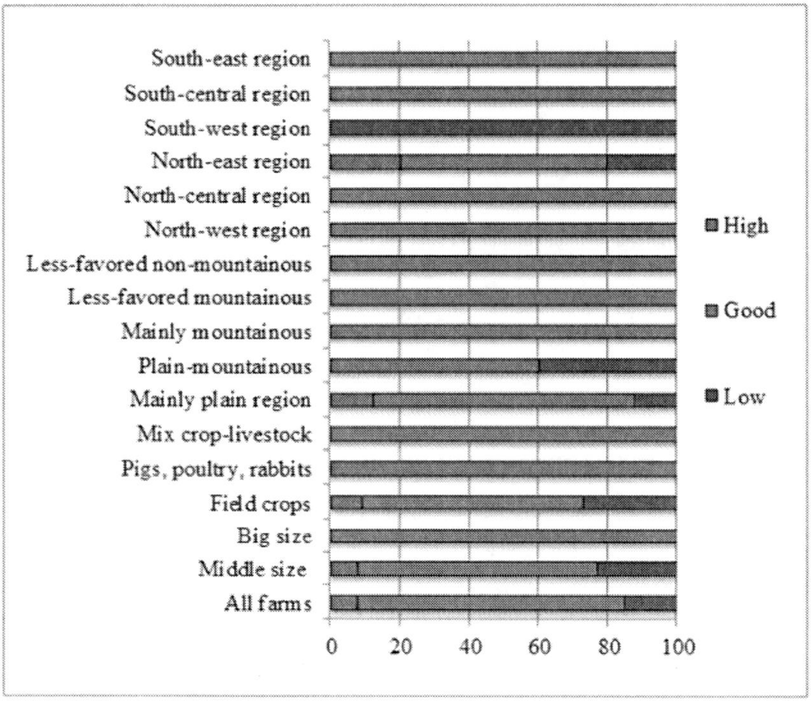

Source: Farm manager survey, July 2016.

Figure 15. Cooperative Enterprises of Various Types with Different Sustainability Levels in Bulgaria (percent).

Cooperative enterprises located in the southwest region had the lowest sustainability. The levels of sustainability of most cooperatives of different types deviated considerably from average levels for sustainability. This suggests that the specific cooperative form (the juridical status of the cooperative) is a critical factor in determining the sustainability of cooperatives of particular types, rather than their belonging to a certain category of holding.

There was significant variation in the number of cooperatives with different sustainability levels for diverse individual types of farming enterprise (Figure 15). All cooperatives of large size; those specializing in pigs, poultry, and rabbits, and crops and livestock; those located in mountainous regions, mountainous and non-mountainous regions with natural handicaps; and those in the northwest, north-central, south-central and southeast regions had good sustainability. The largest proportion of highly sustainable cooperatives were located in the northeast region and in plain regions, as well as those specializing in field crops.

At the same time, cooperatives in the southwest region and 40% of those located in plain-mountainous regions had low sustainability.

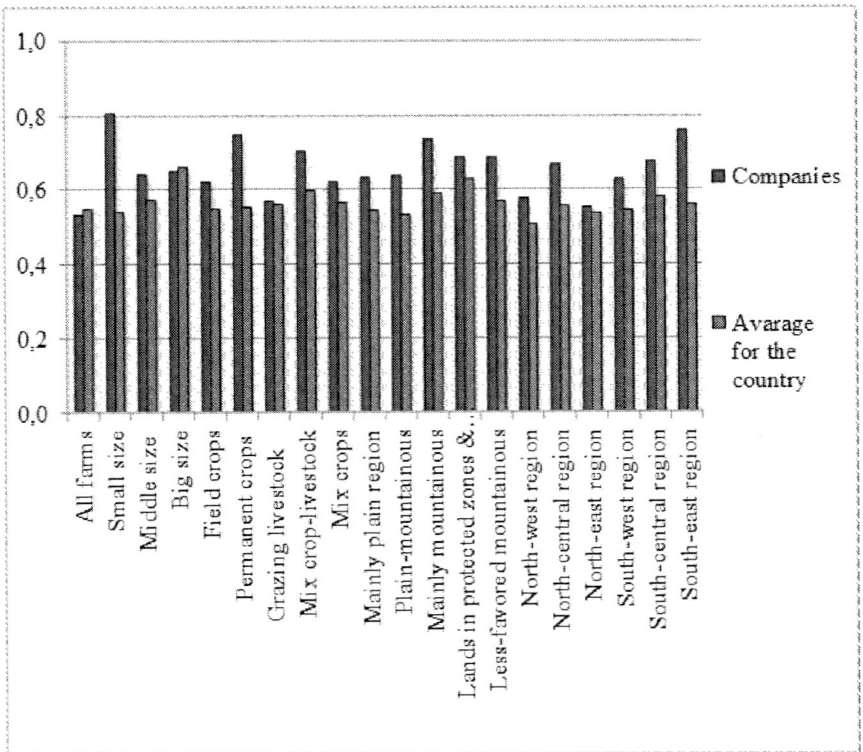

Source: farm manager survey, July 2016.

Figure 16. Levels of Sustainability of Companies of Different Types in Bulgaria.

There was significant variation in the sustainability levels of companies of different sizes, specializations, and locations (Figure 16). The highest levels of sustainability were shown by companies of small size; those specializing in permanent crops; those located in mountainous regions; and those in the southeast region of the country. Companies specializing in grazing livestock and located in the northwest region had lower levels of sustainability. High sustainability was seen in companies of all types, with the exception of firms of large size, specializing in grazing livestock, and located in the northeast region of Bulgaria. This suggests that for most categories of company, their specific juridical status is critical to their level of sustainability. Exceptions include those already mentioned.

Among these companies, there was also significant differentiation between groups of enterprises at different levels of sustainability (Figure 17). All farming enterprises specializing in crops and livestock and those located in mountainous regions with natural handicaps, as well as the vast majority of those of large size for their sector and specializing in mixed crops were highly sustainable. At the same time, half of the companies located in the northwest region of the country and a third of those in the southwest region had low sustainability.

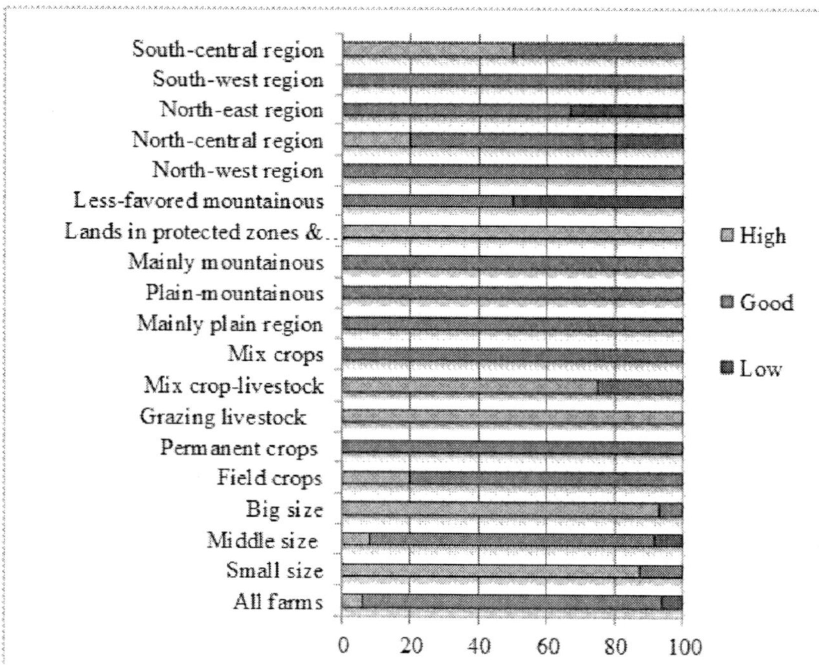

Source: Farm manager survey, July 2016.

Figure 17. Structure of Companies of Various Types with Different Levels of Sustainability in Bulgaria (percent).

FACTORS OF FARMING ENTERPRISE SUSTAINABILITY

Diverse social, economic, market-related, ideological, and personal factors stimulate or restrict the activities of farming enterprises in terms of sustainable operation and development. According to the managers surveyed, factors that could increase the quality of enterprise governance include: available information; official regulations, standards, and norms; access to advisory services; professional training for managers and hired labor; personal convictions and satisfaction; the positive experiences of other farms; innovation; financial capability; private contracts and agreements; and the registration and certification of products and services. For a smaller number of enterprises, factors that could have a role to play in increasing governance include: immediate benefits for other people and groups; regional community initiatives and pressure; existing regional problems and risks; problems and risk existing at the global scale; and problems and risks existing in the country.

Factors encouraging farming enterprises to improve economic sustainability include: market demand and price; direct state subsidies; market competition; financial capability; participation in public support programs; possibility of benefitting immediately; possibility of benefitting in the near future; tax preferences; possibility of benefitting in the long term; and integration with buyers of farm products. Factors considered critical by a smaller proportion of enterprises include: regional community initiatives and pressure; social recognition of individual contribution; pressure and initiatives of interest groups; immediate benefits for other people and groups; and professional training for managers and hired labor.

Factors encouraging the enhancement of social sustainability for the greatest number of farming enterprises include: personal convictions and satisfaction; social recognition of individual contribution; immediate benefits for other people and groups; regional community initiatives and pressure; access to advisory services; European Union policy; and existing regional problems and risks. For a small number of enterprises, important factors encouraging social sustainability include: state control and sanctions; existence of long-term contracts with the state; registration and certification of products and services; tax preferences; and integration with suppliers.

Factors encouraging environmental sustainability include: problems and risks existing at the global scale; official regulations, standards, and norms; existing regional problems and risks; and European Union policies. Significant factors encouraging ecological sustainability for a small number of enterprises include: integration with suppliers; tax preferences; existence of long-term contracts with the state; market demand and price; integration with buyers; market competition; initiatives and pressure from interest groups; partners available for cooperative activities; initiatives of other farmers; and the possibility of garnering immediate benefits.

These incentives need to be examined in relation to the modernization of public policy and the establishment of programs for sustainable development for Bulgarian farming enterprises. Public support through various state and EU CAP mechanisms and instruments seeks to improve multiple aspects of sustainability among farming enterprises in Bulgaria.

This survey has found that current public policies only weakly affect the quality of governance among Bulgarian farming enterprises. National and European Union mechanisms of regulation and support that successfully increase governance in the surveyed enterprises include: professional training and advice; obligatory standards, norms, rules, and restrictions; modernization of agricultural holdings; and establishing produce organizations. Factors impacting governance among a small number of farms include measures such as: afforestation and restoration of forest; natural handicap payments to farmers in non-mountain areas; payments for Natura 2000; and the restoration and development of residential areas.

There are diverse mechanisms of public support that can help improve the economic sustainability of farming enterprises in Bulgaria. Instruments that could impact the economic sustainability of the majority of the surveyed enterprises include: direct area-based payments; national top-ups for products and livestock; modernization of agricultural holdings; green payments; support for semi-market farms. Measures that could considerably improve the economic sustainability of a small number of holdings include: afforestation and restoration of forest; restoration and development of residential areas; stimulation of rural tourism; and the provision of services to residents of rural areas.

The impact that national and European policies have on the social and environmental sustainability of Bulgarian farming enterprises is relatively weak. Instruments that could augment the social sustainability of the majority of farming enterprises include: strategies for local development; the provision of services to residents of rural areas; restoration and development of residential areas; and stimulation of rural tourism. The social sustainability of a small number of holdings could be improved by ecological measures such as: payments for Natura 2000; agricultural environmental payments; and greater support for organic farming.

The most important actions to improve the environmental sustainability of farming enterprises include: green payments; support for organic farming; obligatory standards, norms, rules, and restrictions; and agro-environmental payments. Public instruments that would have the least impact on ecological sustainability of Bulgarian farming enterprises at the current stage of development include: support for setting up micro-enterprises; establishing produce organizations; support for semi-market farms; diversification into non-agricultural activities; support for young farmers; and restoration and development of residential areas

There is a difference shown between individual instruments of public policy and their impact on the sustainability of farming enterprises of different types and locations. Mechanisms and instruments of national and European policy with the greatest impact in improving the sustainability of Bulgarian farming enterprises include:

1) Obligatory standards, norms, rules, and restrictions in terms of the governance of big enterprises and the environmental sustainability of enterprises specializing in pigs, poultry, and rabbits. 2) Direct area-based payments to improve the economic sustainability of: sole traders, cooperatives, companies, holdings of small size for their sector; enterprises specializing in pigs, poultry, and rabbits, mixed crops, and permanent crops; and enterprises located in non-mountainous regions with natural handicaps, those with land in protected zones and territories, the majority

of those in mountainous regions, mountainous regions with natural handicaps, and those in the southwest and south-central regions of the country. 3) National top-ups for products and livestock to improve the economic sustainability of: companies, holdings predominantly for subsistence, and those specializing in grazing livestock; the majority of those in mountainous regions, those with land in protected zones and territories, and those located in the north-central and southwest regions of the country; 4) Green payments to improve the economic sustainability of enterprises located in mountainous regions, those with land in protected zones and territories, and those in the southwest region of the country. 5) Professional training and advice for large enterprises. 6) The modernization of agricultural holdings to improve the economic sustainability of: sole traders and companies; those specializing in mixed livestock and mixed crops; and those located in mountainous regions and in the north-central and south-central regions. 7) Support for semi-market farms and the establishment of produce organizations to improve the economic sustainability of holdings predominantly for subsistence. 8) Natural handicap payments to farmers in mountainous areas to improve the economic sustainability of farming enterprises located in such areas.

All these data on the real impact that individual mechanisms and instruments of public support have on different aspects of sustainability among Bulgarian farming enterprises need to be taken into account when seeking to improve policies and programs supporting agricultural sectors and enterprises of diverse types and locations.

Analysis of the relationships between the personal characteristics of the farming enterprise managers (such as age, gender, competency on sustainability issues) and the level of sustainability of enterprises, allows us to develop a profile of the manager of a sustainable farming enterprise in Bulgaria. More male managers were responsible for holdings with good or high sustainability than female managers. Nevertheless, the high sustainability levels for both genders indicate that there is no significant gender difference in terms of sustainable management.

There also exists a strong correlation between the age of managers and the sustainability of farming enterprises—the largest proportion of holdings with a superior level of sustainability had managers above 65 or younger than 40. A smaller share of managers between 56 and 65 had holdings with good or high sustainability. This suggests that this latter group are more focused on the pure economic vitality of their enterprises (a strategy for profit or survival), or they were not interested in the long-term sustainability of their farming enterprise (planning to exit farming, lack of an heir prepared to carry on farming, etc.).

Estimating the connections between the sustainability of farming enterprises and problems in the region where the holding is located, demonstrates that they are not important—there exists no significant difference in the share of the surveyed holdings with good or high sustainability between regions with different social, economic, and environmental problems. Therefore, the sustainability of farming enterprises appears to depend primarily on managerial capability and strategy, as well as other important external factors (such as public policy) rather than on the specific socioeconomic and environmental challenges in the region where a farming enterprise is located.

There is a strong correlation between the competency of a farming enterprise manager, the extent to which the principles of good governance, and economic, social and environmental sustainability are respected, and the sustainability of a farming enterprise. The proportion of farms with good or high sustainability is greatest, in all aspects, among those that understand and implement principles of sustainable agriculture. Therefore, increasing competency, and developing a culture and practice of sustainable farming is crucial to improving the sustainability of agricultural enterprises in Bulgaria.

The largest share of holdings with good or high sustainability was seen among enterprises with a longer period of existence and a longer period of implementing actions for improving sustainability with maximum values for holdings with a period of between 11 and 15 years. This suggests that sustainable farming requires long-term strategies and targeted action focused on individual aspects of sustainability. A smaller fraction of holdings operating over a period of more than 15 years had good or high sustainability; this is probably a consequence of a lack of effective modernization and adaptation to the constantly changing socioeconomic, institutional, and natural environment over the years.

This analysis highlights the policy instruments that have increased good governance, and economic, social, and environmental sustainability to the greatest extent for the largest number of farming enterprises.

Political mechanisms and instruments that have actually improved the sustainability of Bulgarian farming enterprises include: support for organic farming in respect of social and governance sustainability; adding value to agricultural and forest products in terms of governance; diversification to non-agricultural activities for governance and environmental sustainability; natural handicap payments to farmers in mountain areas, agro-environmental payments, and natural handicap payments to farmers in non-mountain areas to increase social sustainability; and national top-ups for products and livestock to improve governance.

CONCLUSION

This is the first large-scale study on the sustainability of governing structures in Bulgarian agriculture and let us make some important conclusions about the sustainability of holdings in Bulgaria and recommendations for their management and assessment.

The holistic framework suggested offers the possibility of improving the quality of assessment, analysis, and management of sustainability in individual farming enterprises and holdings of different types. This includes important aspects, principles, criteria, and indicators of their governance, and economic, social, and environmental sustainability. New perspectives on governance, relevant indicators, and data drawn directly from farming enterprises have been incorporated into a hierarchical framework offering a comprehensive analytical approach that is easy to understand and use in everyday managerial practice. This novel approach needs to be the subject of further discussion, experimentation, adaptation to specific operational conditions, and improvement to make it appropriate to the study of farming enterprises of different types, subsectors of production, geographical regions, and ecosystems, as well as to meet the specific needs of decision-makers at various levels.

The overall sustainability of Bulgarian farming enterprises is at a good level, with superior levels of environmental and social sustainability, but borderline low levels of governance and economic sustainability. As such, improvement these two aspects is critical to maintaining the sustainability of Bulgarian farming enterprises. Analysis of the primary sustainability principles, criteria, and indicators allows us to identify those components that contribute to the level of each particular aspect of a farming enterprise's sustainability. The levels of governance and economic sustainability of Bulgarian farming enterprises are low because the efficiency of governance and the financial stability of these holdings are low. This is further determined by the low comparative efficiency of the supply of short-term inputs in relation to alternative organizations, the unsatisfactory profitability of the internal capital of enterprises, and the low overall liquidity of farms. Despite overall environmental sustainability appearing relatively high, the preservation of agricultural lands and biodiversity are both relatively low—these are critical to maintaining a satisfactory environmental level. The insufficient application of recommended irrigation norms, high level of soil erosion by water, and reduced numbers of wild animals on farm territory all highlight environmental weaknesses among these enterprises.

A low value for a particular sustainability indicator identifies a specific area requiring improvement through changes in management strategy and/or public policy in relations to farming structures. For instance, despite the overall adaptability of farms being relatively high, the preparedness of farms to adapt to changes in the natural environment (climate, extreme events, etc.) is relatively low. Unless effective measures are put in place to improve this adaptability through education, training, information, improved agro-techniques, new production structures, new varieties, and technological and organizational innovation, the long-term sustainability of Bulgarian farming enterprises will be compromised. The high values for certain sustainability indicators show the areas where Bulgarian farming enterprises have significant absolute and comparative advantages for sustainable development. At the current stage of development, these include: respecting animal welfare standards; protection of surface and ground water from contamination with nitrates and pesticides; preservation of air quality; implementation of good agricultural practices; limited number of livestock per unit of farmland; acceptable labor conditions; comparative satisfaction gained from farming activity; optimal productivity of livestock; good adaptability to the market (price, competition, demand); and comparative governance efficiency of the marketing of products and services.

There is great variation in the sustainability of farming enterprises of different types and locations. The best overall sustainability is shown by: companies, cooperatives, and enterprises of large size; holdings specializing in pigs, poultry, and rabbits; those with land in protected zones and territories; those located in non- mountainous regions with natural handicaps; and those in the south-central region of the country. Holdings predominantly for subsistence and those specializing in mixed livestock have low sustainability. Those enterprises specializing in vegetables, flowers, and mushrooms, physical person enterprises, and enterprises located in the northwest region of the country have good levels of sustainability, but with some areas being borderline. For these enterprises, effective measures have to be put in place to improve their sustainability.

Farming enterprises of small size; those specializing in mixed crops and permanent crops; those located in plain-mountainous regions, mountainous regions; and those in the northeast and southwest regions of the country have low economic sustainability. For these types of enterprise, effective action needs to be undertaken to increase economic sustainability and improve their overall long-term sustainability.

Sole trader farming enterprises have low social sustainability and effective measures need to be introduced for the improvement of this aspect of sustainability, such as training, regulation, and support.

There is significant variation in the values of particular sustainability indicators for particular farming enterprises and those of different juridical types, sizes, specializations, and geographical locations. Specific analysis of the latter also indicates areas where adequate measures need to be put in place by enterprise managers and the state to improve specific aspects and the overall sustainability of farming enterprises of different types.

There is great variation in the share of farming enterprises with different sustainability levels: more than 68% of Bulgarian farms have good sustainability and less than 2% have high sustainability. At the same time, one quarter of farming enterprises in the country have low sustainability and a little more than 4% are unsustainable. Companies make up the greatest share of farming enterprises with good or high sustainability, followed by cooperatives and sole traders. More than a third of physical person enterprises have low sustainability or are unsustainable. One quarter of sole traders have low overall sustainability, as do 15% of cooperatives; only 6% of companies have low sustainability. All farming enterprises of large size have good sustainability. More than a half of holdings predominantly for subsistence have low sustainability or are unsustainable. Similarly, around a third of enterprises of small size and 24% of those of medium size have low sustainability. The proportion of enterprises with good or high sustainability is greatest among those specializing in: pigs, poultry, and rabbits; mixed crops; permanent crops; mixed crops and livestock; field crops; and grazing livestock. On the other hand, 43% of enterprises with mixed livestock have low sustainability and 14% are unsustainable; similarly, 41% of enterprises specializing in vegetables, flowers, and mushrooms have low sustainability and 5% are unsustainable. The share of farming enterprises with good or high sustainability is most significant among those located in non-mountainous regions with natural handicaps; those with land in protected zones and territories; those in plain regions; and those in the south-central, north-central, and southeast regions of the country. 40% of all enterprises in the southwest region have low sustainability or are unsustainable; 37% of those in the northwest region and 32% in the northeast region also have low sustainability. The northwest region has the highest number of unsustainable farming enterprises—a tenth is unsustainable. A huge number of holdings in mountainous regions with natural handicaps (38%), 35% of those located in mountainous regions, and a third of those in plain-mountainous regions have low overall sustainability or are unsustainable.

Around 40% of farming enterprises have low governance or are managerially unsustainable. More than 42% of all farming enterprises in the country have low economic sustainability or are economically unsustainable. 18% of farming enterprises have low social sustainability and one tenth is socially unsustainable. 18% of farming enterprises have low environmental sustainability and a little more than 4% are ecologically unsustainable.

There exists considerable variation in the sustainability of physical person enterprises, sole traders, cooperatives, and companies according to their size, specialization, ecology, and geographical location. There are also significant differences in the shares of enterprises with different levels of sustainability among diverse kinds of enterprises. For physical person enterprises, size, product specialization, and location are more important than their juridical status. For certain kinds of sole trader, cooperative, and company, the specific organizational form is a crucial factor determining sustainability, rather than their belonging to group of a particular size, specialization, or location.

Factors encouraging the improvement of governance among Bulgarian farming enterprises include: access to advisory services; professional training of managers and hired labor; personal convictions and satisfaction; the positive experience of other farms; innovation; financial capability; private contracts and agreements; and the registration and certification of products and services. Factors that encourage farming enterprises to improve economic sustainability include: market demand and price; direct state subsidies; market competition; financial capability; participation in public support programs; the possibility of benefitting immediately; the possibility of benefitting in the near future; tax preferences; the possibility of benefitting in the long term; and closer integration with buyers. Factors that encourage enhanced social sustainability include: personal convictions and satisfaction; social recognition of individual contributions; immediate benefits for other people and groups; regional community initiatives and pressure; access to advisory services; policies European Union; and existing regional problems and risks. Factors that encourage farming enterprises to increase environmental sustainability include: problems and risks existing at the global scale; official regulations, standards, and norms; existing regional problems and risks; and European Union policies. All these incentives have to be taken into account in planning the modernization of public policy and programs for sustainable development.

National and European mechanisms of regulation and support that affect the economic sustainability of the majority of Bulgarian farming enterprises include: direct area-based payments; national top-ups for products and livestock; modernization of agricultural holdings; green payments; and direct support for semi-market farms. The impact of national and European policies on the quality of governance and the social and environmental sustainability of Bulgarian farming enterprises is relatively weak. There are large differences in the impact of individual policy instruments on the sustainability of farming enterprises of different types and locations.

Keeping in mind the importance of the holistic assessment of farming enterprise sustainability and the enormous benefits that this provides for farm management and agrarian policy making, this type of study needs to be extended and expanded in scope. This requires close cooperation between all interested parties and the participation of farmers, agrarian organizations, interest groups, research institutes, experts, and local and state authorities. Moreover, the precision of estimates has to be improved through the incorporation of other relevant information from field tests and surveys, the expertise of relevant professionals, and statistical and other data into the process.

References

Andreoli Maria and Tellarini Vittorio. (2000). Farm sustainability evaluation: methodology and practice, *Agriculture, Ecosystems & Environment*, Volume 77, Issues 1-2 43-52.

Bachev, Hrabrin. (2005). Assessment of Sustainability of Bulgarian Farms, proceedings, *11th Congress of the European Association of Agricultural Economists*, Copenhagen.

Bachev, Hrabrin. (2010). *Governance of Agrarian Sustainability*, New York: Nova Science Publishers. 116.

Bachev, Hrabrin. (2014). *Environmental Management in Agriculture, Mechanisms, Forms and Efficiency*, LAP LAMBERT Academic Publishing.

Bachev, Hrabrin. (2016). A Framework for Assessing Sustainability of Farming Enterprises, *Journal of Applied Economic Sciences*, Vol. XI, 1(39): 24-43.

Bachev, Hrabrin. (2016). Sustainability of Farming Enterprise—Understanding, Governance, Evaluation, *EKOHOMIKA*, 2, 179: 6-15.

Bachev, Hrabrin. 2016. Unpacking Sustainability of Farming Organizations, *International Journal of Economics and Management Sciences*: 5:3, 1-13.

Bashev, Hrabrin. 2016. The Sustainability of Farms, *Economics*, no 1: 20-53.

Bachev, Hrabrin and Tsuji, Masao. (2001). Structures for Organization of Transactions in Bulgarian Agriculture. *Journal of the Faculty of Agriculture of Kyushu University*, 46 (1): 123-151.

Bachev, Hrabrin and Peeters, Alain. (2005). Framework for Assessing Sustainability of Farms, in *Farm Management and Rural Planning* No 6, Kyushu University, Fukuoka: 221-239.

Bachev, Hrabrin and Nanseki, Teruaki. (2008). Environmental Management in Bulgarian Agriculture—Risks, Modes, Major Challenges. *Journal of the Faculty of Agriculture of Kyushu University*, 53 (1): 363-373.

Bachev Hrabrin, Ivanov Bodjidar, Toteva Dessislava, Sokolova Emilia. (2016). Agrarian Sustainability and its Governance—Understanding, Evaluation, Improvement, *Journal of Environmental Management and Tourism*, Vol. 7, 4: 639-663.

Bastianoni Simone, Marchettini Nadia, Panzieri Margherita, Tiezzi Enzo. (2001). Sustainability assessment of a farm in the Chianti area (Italy), *Journal of Cleaner Production*, Volume 9, Issue 4, 365-373.

Brklacich Michael, Bryant Christopher and Smith Barry. (1991). Review and appraisal of concept of sustainable food production systems, *Environmental Management*, 15(1): 1-14.

EC. (2001). *A Framework for Indicators for the Economic and Social Dimensions of Sustainable Agriculture and Rural Development*, European Commission,

Edwards Clive, Lal Rattan, Madden Patric, Miller Robert and House, Gar (editors) (1990). *Sustainable Agricultural Systems*, Soil and Water Conservation Society, Iowa.

FAO: SAFA. (2013). *Sustainability Assessment of Food and Agriculture systems indicators*, FAO.

Fuentes, Maria. Farms Management Indicators Related to the Policy Dimension in the European Union, *OECD Expert Meeting on Farm Management Indicators and the Environment*, 8-12 March 2004, New Zealand.

Häni Fritz, Pintér Laszio and Herren Hans. Sustainable Agriculture. From Common Principles to Common Practice, *Proceedings of the first Symposium of the International Forum on Assessing Sustainability in Agriculture (INFASA)*, March 16, 2006, Bern, Switzerland.

Hansen, James. (1996). *Is Agricultural Sustainability a Useful Concept, Agricultural Systems*. 50: 117-143.

Hayati Dariush, Ranjbar Zahra Karami Ezatollah. (2010). Measuring Agricultural Sustainability, in E. Lichtfouse (ed.), Biodiversity, Biofuels, Agroforestry and Conservation Agriculture, 73, *Sustainable Agriculture Reviews* 5, Springer Science+Business Media B. V., 73-100.

Hendricks, Nathan. (2010). *The Effect of Green Payments on the Diffusion of Conservation Technologies*.

Garnett, Tara and Godfray, Charles (2012). *Sustainable intensification in agriculture. Navigating a course through competing food system priorities*, Food Climate Research Network and the Oxford Martin.

Gesinaldo Candido, Nobrega Mariana, Figueiredo Marilia Maior Monica. *Sustainability Assessment of Agroecological Production Units: A*

Comparative Study OF IDEA and MESMIS Methods, Ambiente & Sociedade, Ambient. soc. vol. 18 no.

Ikerd, John. (2016). *On Defining Sustainable Agriculture*, SARE. http://www.sustainable-ag.ncsu.edu/onsustaibableag.htm

Lowrance Richard. (2015). Hendrix Paul and Odum Eugene: A hierarchical approach to sustainable agriculture, *American Journal of Alternative Agriculture*

Lopez-Ridaura Santiago, Masera Omar, and Astier Marta. (2002). *Evaluating the Sustainability of Complex Socio-environmental Systems*. MESMIS Framework, Ecological Indicators.

Mirovitskaya, Natalia and Ascher, William, (editors) (2001). *Guide to Sustainable Development and Environmental Policy*, Duke University Press, London.

OECD. (2001). *Environmental indicators for agriculture. Volume 3: Methods and Results*. OECD, Paris.

Raman, Saroja. (2006). *Agricultural Sustainability. Principles, Processes and Prospect*, New York: The Haworth Press Inc.

Rigby, Dan, Woodhouse, Phil, Young, Trevor, and Burton, Michael. (2001). Constructing a farm level indicator of sustainable agricultural practice, *Ecological Economics*, Vol. 39, Issue 3: 463–478.

Sauvenier Xavier, Valekx Jan, Van Cauwenbergh Nora, Wauters Erwin, Hrabrin Bachev, K. Biala, Bielders Charles, V. Brouckaert, V. Garcia-Cidad, S. Goyens, Hermy Martin, Mathijs Erik, Muys Bart, Vanclooster Marnik and Peeters Alain. (2005). *Framework for Assessing Sustainability Levels in Belgium Agricultural Systems*—SAFE, Belgium Science Policy, Brussels.

UN. *Report of the United Nations Conference on Environment and Development*, 3-14 June 1992, Rio de Janeiro: United Nation.

UN. *Paris Climate Change Conference*, November-December 2015.

UN. *Transforming our world: the 2030 Agenda for Sustainable Development*, United Nations Resolution A/RES/70/1 of 25 September 2015.

Van Loon, Gary, Patil, S. G., and Hugar, L. B. (2005). *Agricultural Sustainability: Strategies for Assessment*. London: SAGE Publications 280.

Williamson, Oliver (1996). *The Mechanisms of Governance*. New York: Oxford University Press, 165.

In: Agribusiness
Editor: Ronald Luna
ISBN: 978-1-53614-413-0

Chapter 2

NANOTECHNOLOGIES: MARKET STRUCTURING AND COMPETITIVE STRATEGIES OF AGRIBUSINESS MARKETS

Vincent Sabourin*
Department of Strategy and Social Responsibility ESG School of Management, University of Quebec at Montreal, Canada

ABSTRACT

The rising demand for fast speed has generated the need for nanotech based systems and technologies, thus driving the growth of the nanotechnology markets. Nanotech finds application in different fields such as medical technology, production technology, lighting components and defense nanotechnics. Agribusiness sector is poised to be a 3 trillion-dollar industry in global investment by 2030 (World Bank 2013). Nanotechnology has the potential to impact significantly on all sectors of agribusiness ecosystem in the next decade. Nanotechnology could be used to enhance the possibilities of re-developing conventional and stranded agribusiness resources. Nanotechnology can make the industry considerably greener and competitive, with its current growth rate of 30% (US$ 1.08 billion) annually. Nanotechnology, holding a focus focusing

* Corresponding Author Email: sabourin.vincent@uqam.ca.

on unique attributes of materials coming through from smaller sizes closer to nanometric, has the vibrancy to disrupt the agribusiness sector with an extension to other fields such as environmental engineering, biomedicine, defense and security, water utilization and conversion of energy. It is well noted that adoption of new technology and innovation is crucial in accumulation of global wealth and market value which now stand over US$ 1.2 trillion in a rough estimate. Nanotechnology has positioned itself as an advancement in technology that has the potential to create and revolutionize the whole agribusiness chain, with the ability to raise the bar for agricultural yield and efficiency, thus ensuring that there is sustainable food security and economic gain for key industrial players by 25% over and above the industry growth rate. It examines the strategies of adoption which companies use to obtain market shares. The structuring of the market under various market applications is also examined. Finally, the article looks at the positioning of the players with respect to economies of scale and competitive advantages.

Keywords: agribusiness, nanotechnology, agribusiness sector, competitive analysis, market growth

INTRODUCTION

Nanotechnology, which falls among new enabling technologies, has the potential and capacity to revolutionize agriculture and food systems globally (European Commission, 2013). Agricultural and food systems, disease management delivery systems, modern appliances for molecular and cellular biology, modern materials for pathogen diagnosis and detection and environmental protection are classical examples of the vital potential of nanotechnology to the study and revolution of agriculture and food systems globally (Dlamini et al., 2014; Das et al., 2013; De Azeredo et al., 2013). According to European Nano Business Association (2005), in-depth market analysis of these technologies as well as trends, forecasts and profiles of major players indicate how valuable, the growth of nanotechnology has become. Hurmelima, Kylaheiko and Jauhiainen (2007) further argue that efficiency of nanotechnology has led to great discoveries in prescription drug products. Agriculture has for a long time, dealt with the challenge of improving efficiency in crop productivity, processing of food, ensuring safety of food and equally, the management of the consequences of producing, storing and distributing food on the environment (Banterle et al., 2014; Sekhon 2014). Wilson et al., (2014) infer that nanotechnology is a part of any nation's future stability. Quite evidently, this research has an extremely great potential to benefit society through technological applications in agriculture and food systems. Agriculture remains to be the backbone of most developing economies, with more than 60% of the population relying on it to derive their livelihood (European Commission, 2013; European Food Society Association, 2014). Nanotech technologies play an increasingly important role in various markets of the global economy. However, various technical, marketing and other hurdles need to be overcome before Nanotech's can realize their full potential (Wu, 2004). Although Nanotech-enabled technologies are growing rapidly, the shortage of skilled manpower is inhibiting the growth of this field. Nanotechnology can significantly improve our comprehension of the biology of different crops and thus potentially enhancing yield production or nutritional values (Wilson et al., 2014; Khodos, 2014). However, this is only possible through the development of improved systems for monitoring environmental conditions and delivering of appropriate nutrients or pesticides that are environmentally friendly (De Azeredo et al., 2013; Okechukwu, 2014; Wilson et al., 2014; Khodos, 2014). In addition, Lopez-Salazar et al., (2014) have argued that nanotechnology can buffer up opportunities for value added crops or environmental remediation. Globally, many economies have discovered the potential of nanotechnology in the agribusiness sector and are fast investing significant amounts in it (Banterle et al., 2014; Hirsch et al., 2014). Equal importance has also been given to societal issues associated with nanotechnology and to improve agribusiness awareness (De Azeredo et al., 2013). Duran and Marcato (2013 have established that demand for advanced agricultural products is increasing year on year due to their value addition potential, to target end users. Apart from this, many new technological applications have emerged due to commercialization of manufacturing techniques within the agribusiness sector (Kafi and Ghomi, 2014). Agribusiness sector has massive potential not just now but in the future, as it would create range of applications, namely in key industries such as food processing sector, animal husbandry, plant enculturement among others (EFSA, 2011; Kuan and Ye-Found, 2012; Duran and Marcato, 2013). Hutchison (2008) argues that end-market demand for nanotechnology is unclear, especially since there are only a limited number of commercial grade products that can be compared to conventional materials in terms of performance and market success, and the applications are often just as innovative as the materials themselves. Of these challenges, one important common feature is that many of them are the result of issues that occur well in advance of commercialization, i.e., during the design and production process development phases (Moraru, 2003; Bonaccorsi and Thoma, 2005). Improvements in specific characterization and data analysis tools would have an impact on these issues. Most analyses of the Nanotechs market focus on the development of new Nanotechs technologies and applications, rather than assessing the potential market for these technologies (Chaudhry, 2008; Cohen, Goto, Nagata, Nelson and Walsh, 2002). Several analyses have made valuable contributions to raising investors' awareness and interest in Nanotechs (Wonglimpiyarat, 2005). However, these analyses by themselves do not provide sufficient information to guide corporate or individual investment decisions. Investors require additional information, such as the size of specific Nanotechs markets, prices, competition, and potential regulation (European Commission, 2004a).

Thus, a comprehension of nanotechnologies in agribusiness markets and industry would offer intelligence on the entire value chain and market segmentation and this would enable key industry stakeholders to make strategic as well as detail oriented decisions on agribusiness investments.

Objectives of the Article

This article consisted of both general and specific. The main objective of this article was to study the market dynamics and categorize it into five competitive strategies in the Nanotechs markets:

- To study the new possibilities of Nanotechnologies in agribusiness sector;
- To study the growth and market trends of nanotechnologies in agribusiness sector;
- To assess the adoption strategies of Nanotechnologies in agribusiness sector;
- To identify market segmentation of Nanotechnologies in agribusiness sector;
- To assess the competitive position of Nanotechnologies in agribusiness sector.

NEW POSSIBILITIES OF NANOTECHNOLOGY IN AGRIBUSINESS SECTOR

The increase in human population and demand for habitat, utilizes agriculture land and water resources leading to depletion of key soil nutrients (Bucheli, Knauer and Gogos, 2013). A report by Duracio et al., (2013) highlighted that a large proportion of those living in developing economies are facing daily food shortages while in the developed world there is food surplus. European Commission (2013) had earlier reported that to developing economies, the drive is to develop drought and pest resistant crops, which also maximizes on yield production. In developed countries, the agro-food industry is driven by consumer demand for fresher and healthier foodstuffs (European Food Safety Authority, 2012). Worrying trends in population growth indicate that food production in the near future will not match the ever increasing world population (Dlamini et al., 2014). In this regard, Banterle et al., (2014) infer that is through innovative agriculture that the world can envisage a self-sustainable agribusiness sector. With limited land and water resources, demand-output ratio in agriculture can be met only by increasing productivity through the effective use of modern technology. Nanotechnology, focusing on special properties of materials emerging from nanometric size has the potential to revolutionize the agricultural and food sectors, biomedicine, environmental engineering, safety and security, water resources, energy conversion, and numerous other areas (Hirsch et al., 2014; Dlamini et al., 2014). In fact several products enabled by nanotechnology are already in the market, such as antibacterial dressings, transparent sunscreen lotions, stain-resistant fabrics, scratch free paints for cars, and self-cleaning windows (Sastry, Aushul and Rao, 2013).

Wilson et al., (2014) have predicted that nanotechnology will transform the entire food industry, changing the way food is produced, processed, packaged, transported, and consumed. An estimate by market research and industry analysis shows that, the market for the nanotechnology was 7.6 billion US$ in 2008 and was expected to be US$1 trillion in 2030 (Sadano and Vernaeau, 2014). The application of nano-technology in disease control, slow release of pesticides and developing diagnostic tools, and development of functional food systems, to produce interactive, edible nano wrappers to keep the pathogens away, targeted release of chemicals, packaging, extensive nano surveillance, interactive agrochemicals as herbicides and pesticides (Lopez-Salazar et al., 2014; Brzozowska, 2014). In terms of emerging trends in global agribusiness sector, Aquaculture is the world's fast growing area of animal production, and it is projected to be one of the first industries to integrate and commercialize nanotechnology products (Duran and Marcato, 2013). Potential applications include nanodelivery of veterinary products in fish food, antibacterial surfaces in the aquaculture system, and nanosensors for detecting pathogens in the water. Although potentially beneficial, there are existing knowledge gaps regarding the impacts of nanoparticles on aquatic organisms (Sekhon, 2014). In plant-based agriculture, emerging applications include nanoformulated agrochemicals (e.g., fertilisers, pesticides, biocides and veterinary medicines) for improved efficiency, reduced use of farm chemicals, new toxin formulations for pest management, and better control of applications (e.g., slow release of pesticides). For example, nanosensors can be used for the detection of pathogens, pesticides and other chemicals. Nanosensors have been applied in pesticide residue detection such as organophosphate in fruit, plants and water (Duran and Marcato, 2013). Khot et al., (2012) have argued that nanosensors offer high sensitivity, low detection limits, super selectivity, fast responses, and small sizes.

Sources: Syed Baker, Tatiana Volova, Svetlana V. Prudnikova, S. Satish, Nagendra Prasad M. N. Nano-agroparticles emerging trends and future prospect in modern agriculture system. *Environmental Toxicology and Pharmacology*: Vol 53, 10-17.

Figure 1: Nanoparticles: The changing dynamics in agribusiness sector

TRENDS OF NANOTECHNOLOGY IN AGRIBUSINESS SECTOR

Agribusiness is projected to be a $ 2.9 trillion industry in global investment by 2030 (World Bank 2013). It is therefore understandable that agricultural development in different economies is back on global development and investment agenda. With investments from sources around the globe, agribusiness started booming in the 2000s. In addition, soaring grain prices and global food inflation spurred investor interest in farming (Vink, 2010). Many farming projects, linked to the value chain through agri-food processing, food retail establishments and service networks, increasingly do business on the continent, due in part to growing private sector interest (World Bank, 2013). World agriculture attracted more than $1.8 trillion in private equity investment in the first half of 2013, compared with $0.8 trillion for the whole of 2012 (Rema and Martinez, 2013). This trend is also recognized and welcomed in the world of government and economic politics. In September 2009, the G20 recognized agricultural development as a central activity for emerging economies development, echoing calls by the World Development Report (World Bank, 2103). The emerging consensus among development institutions and agencies and strategic investors, referred to in the above paragraphs, has been appreciated as a primary set of forces driving the restructuring of agricultural development in the global economics (Ortega, 2013; Scoones et al., 2014). The universally accepted Millennium Declaration, which emphasizes on agriculture's prominent role in all eight of the MDGs, together with the declarations by world leaders on the future role of agriculture, clearly situate agriculture's roles and contributions within the broader economic-political context of both global and local realities impacting on individual economies (De Janvry, 2009; World Bank, 2013). By 2050, the major growth in demand for food and products is expected to come from the markets of North America, Western Europe and China, and these trends will drive commercial food business systems (Scoones et al., 2014). Demand will be influenced by population growth, per capita income trends, lifestyle aspirations and related consumer preferences. Consumer demand in these regions has become and will continue to be more exacting, fragmented and geared to convenience, food safety and quality (Bairwa et al., 2014). While these consumers will also seek attributes of fun, surprise and taste sensations in their food experiences; they will also be sensitive to environmental, ethical and social considerations (Hughes 2007; Vermeulen et al., 2008). According to Khodos (2014), these trends are expected to have profound effects on the sourcing of food products, and will link farm production, processing, wholesaling and retailing. In support of this view, Raynolds et al., (2014) assert that agri-food value chain will increasingly be subjected to consumer preferences, considerations and values originating in these markets, and the related food safety, environmental and ethical valuations, certifications requirements, traceability and monitoring systems. Another trend that has been witnessed in the agribusiness sector is the changing investment patterns. The restructuring of the agri-food business system is also expected to change the nature of agribusiness investments (Alishahi et al., 2014; Brenes et al., 2014). For instance, Raynolds et al., (2014) have argued that in order to consolidate value chain actions and allow for scale economics, investments will need to focus on wholesale market infrastructure. The singular purpose of providing farm producers with access to the next level in the value chain will thus shift the emphasis to the development of infrastructure and support systems to facilitate the functioning of all levels of operation in the total agri-value chain (Rhim, Park and Ha, 2013; Khodos, 2014). The new drivers of business opportunities in the food system will be investments that focus on agribusiness mentoring, coaching and extension for producers participating in the value chain, and on processing, retail functions, and their mechanisms for support and food safety (Swinnen 2010; Reardon et al. 2009; McKinsey 2011; World Bank 2013). A report by European Commission (2013) showed that smart field sensing systems are increasingly becoming important applications for the real time monitoring of crop growth and field conditions including nutritional status, light, temperature, moisture level, soil fertility, insects, weeds and plant diseases to maximize yields for sustainability in changing climatic conditions. Chen and Yada, (2011) has reported that networks of wireless nanosensors placed across cultivated fields provide detailed information on crop and soil conditions, enabling the best agronomic decisions to be made while minimizing resource inputs. This includes information on the optimal times for planting and harvesting crops, as well as the time and level of water, fertilisers, pesticides, and other treatments that are required to be administered given precise plant physiology, pathology, and environmental conditions (Chen and Yada, 2011). Wireless nanosensors have already been used in certain parts of the U.S. and Australia (Duran and Marcato,

2013). For instance, a Californian vineyard, Pickberry, in Sonoma County has installed Wi-Fi systems with the aid of the information technology company, Accenture. The cost of installing this system is rationalized by the fact that it facilitates the best grapes to be grown which in turn produces better-quality wines, which command a premium price (BCC, 2011).

Market Size of Nanotechnology in Agribusiness Sector

Global economic reports estimated that the global agribusiness market in 2010 ranged from about US$ 20.7 billion to US$ 0.98 trillion. By 2020, the market worth is estimated to be more than US$ 3.4 trillion (BCC, 2011). This underscores the value of raising awareness of nanotechnology in agribusiness market, equipping the sector with the skill sets and technical know-how to handle the special technologies and adhere to the Environment, Health and Safety regulations (Bayer, 2010). In the agri-nano, researchers have highlighted need for technological and engineering issues relating to scalability, quality and reliability of adopting advanced agricultural materials in value addition, as well as the need for closer collaborations amongst the government, academic and industry players to bring down the high cost associated with nano-materials in agribusiness sector. This collaboration will certainly lead to expansion in the projected growth and size of agribusiness market to USD 160 billion by 2020 (see Figure 2 below).

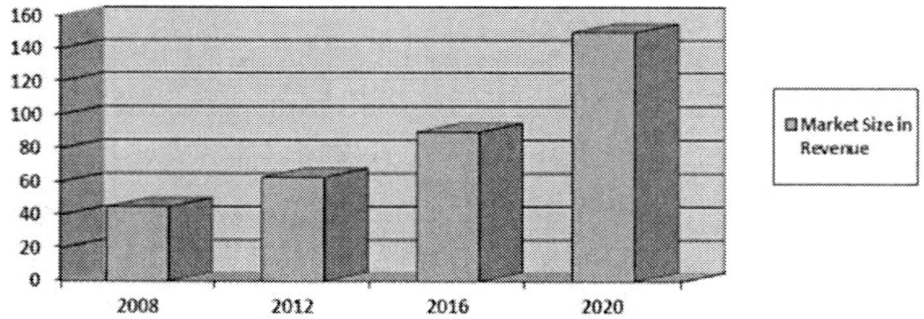

Figure 2. The Projected growth and future size of Agribusiness Sector.

Sourced from BCC Research (2014)

From the findings of Bukucs et al., (2014), nanotechnology applications in agriculture has been described as the new industrial revolution and both developed and developing countries are investing in this technology to secure a market share. At present the USA leads with a 4 year, 3.7 billion USD investments through its National Nanotechnology Initiative (NNI) (Schiefer and Hortmann, 2014; Banterle et al., 2014). The USA is followed by Japan and the European Union, which have both committed substantial funds (USD 750 million and USD 1.2 billion, including individual country contributions, respectively per year) (Sodano and Vernaeau, 2014).Yusuf (2014) notes that the level of funding in developing countries may be comparatively lower, however this has not lessened the impact of some countries on the global stage. A recent study from the Helmuth Kaiser Consultancy predicts that the nano-food market will surge from 2.6 billion USD to 20.4 billion USD by 2018 (Brenes et al., 2014). More than 400 companies around the world today are active in nanotechnology research and development (RandD) and this number is expected to increase to more than 1000 within the next 10 years. In terms of numbers, the USA leads, followed by Japan, China, and the EU. An estimate by the Business Communications Company, a technical market research and industry analysis company shows that, the market for the nanotechnology was 7.6 billion USD in 2003 and is expected to be 1 trillion USD in 2020 (BCC, 2014). However, Wilson et al., (2014) argues that the full potential of nanotechnology in the agricultural and food industry has still not been realized. Nanotechnology will leave no field untouched by its ground breaking scientific innovations (Dlamini et al., 2014). The agriculture and food industry is no exception. Lopez-Salazar et al., (2014) reported that nanotechnology has provided new solutions to problems in animals, plants, plant products and post-harvest technology in enhancing the quality of agricultural products, with a commercial focus. It has a significant effect in the global agribusiness industry - development of new functional materials, product development, and design of methods and instrumentation for food safety and bio-security (Lopez-Salazar et al., 2014). This is estimated to grow by 12.5% for the next 5 years to USD1.9 trillion (Hirsch et al., 2014).

Market Segmentation of Nanotechnology in Agribusiness Sector

The global agribusiness market sector is segmented majorly by type of application. These applications include:

Nanotech Materials in Agricultural Production Segment

During primary production nano-formulated agro-chemicals are employed to increase the efficacy of the agro-chemicals compared to conventional formulations (De Azeredo et al., 2013; Das et al., 2013). For example, nano-encapsulated and solid lipid nanoparticles have been explored for the delivery of agrochemicals (Danisch, 2013). The application of novel nanotechnology techniques in agriculture has recently been reviewed by a number of authors (Gogos et al., 2012: Narayanan et al., 2012: Kushwaha et al., 2012: Shamim, 2011: Mahalik et al., 2010) and can be divided into three main categories; pesticides, plant growth promoters and animal feed additives. Most search results in this category deal with organic nanomaterials, followed by titanium dioxide, silver, silica, and alumina (Shamim, 2011: Mahalik et al., 2010). With relation to application, nanotechnology as an emerging field will find agriculture to be deficient market that requires support in terms of research and development (Cota-Arriola et al., 2013; De Azeredo et al., 2013). This direct entry offers a wide market with unending needs that cannot be met.

Nanotech Materials in Food Processing/Additives Segment

This segment offers that nanotechnologies are applied in food production machinery and in food additives (Cota-Arriola et al., 2013). While direct food contact is evident, this application of nanotechnology is expected to have very low additional safety concerns, because migration or carry-over to food should be excluded and exposure is therefore expected to be negligible (Plascencia-Jatomea, 2013; Cota-Arriola et al., 2013). An exception may be cooking equipment, crockery and other kitchenware that may have been treated with nanomaterials to provide it with antibacterial properties (Chen et al., 2010) as in that case the particles are on the surface to exert their microbial activity. Of more importance are food additives which are substances added to food to improve the stability of foods during processing and storage, to improve certain product characteristics, or to increase the potency and bioavailability of nutrients in the food product (Ali et al., 2014; Khodos et al., 2014).

Nanotech Materials in Food Packaging Segment

In this segment, incorporation of nanomaterials or nanotechnological devices in packaging materials or storage containers in order to lengthen the storage time while keeping the products fresh are considered the most important type of nanotechnology application in the food area for the near future (Chaudhry et al., 2010; Cota-Arriola et al., 2013; Khodos, 2014). Silvestre and Cimmino (2013) have inferred that nanocomposites can improve mechanical strength; reduce weight increase heat resistance and improve barrier against oxygen, carbon dioxide, ultraviolet radiation, moisture, and volatiles of food package materials. Main nanoparticles that have been studied for use in food packaging systems as well as their effects and applications were reviewed recently (Vink, (2010; Papaspyrides, 2010: Silvestre et al., 2013).

A summary of the applications of nanotechnology in agribusiness sector can be deduced in the Figure 3.

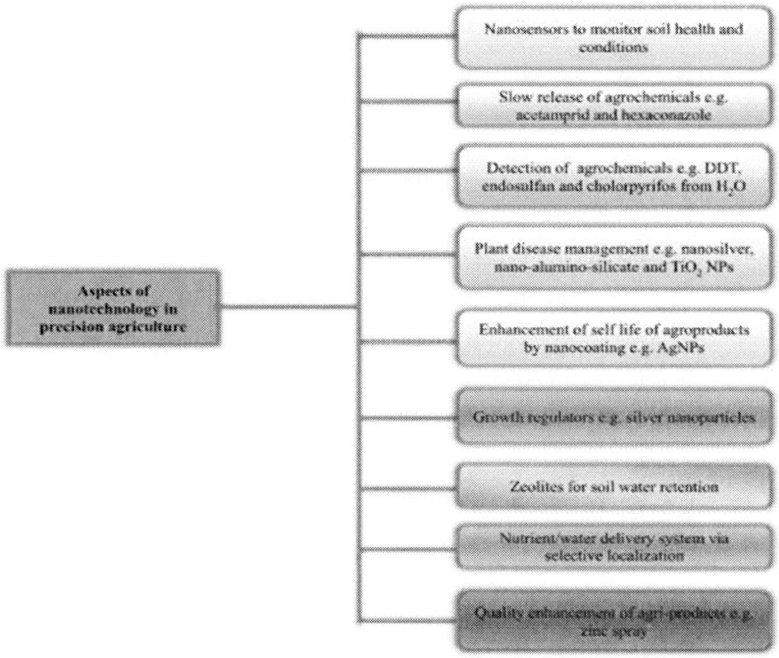

Source: Kaur, Pawan. (2017). Nanotechnology: The new perspective in precision agriculture A R T I C L E I N F O. *Biotechnology Reports*. 15. 10.1016/j.btre.2017.03.002

Figure 3. Applicable scope of Nanotechnology.

Strategies for Adoption of Nanotechnology in Agribusiness Sector

Ranjan et al., (2014) note that a strategy for adoption involves mostly recreating or installing new systems with the goal of offering a less expensive technology to the buyer. This strategy requires important investments and contains its own risks (Bukucs et al., 2014). However, it facilitates the creation of substitution costs by attaching the buyers to the solution technology which is proposed (Scoones et al., 2014; Markets and Markets, 2009). As relates to agribusiness markets, Magnuson et al, (2014) argue that there are collaborative alliances among industry, academia and governments seeking to raise awareness of nanotechnology and its impact on everyday growth of agribusiness industry. Wilson et al., (2014) on the other hand note that there is an increased cooperation and coordination among different agribusiness industries, government and academia to advance Nanotechs-driven fields; and drive world funding and investment in areas of agribusiness is critical to maintaining global economic competitiveness and food security. A report by European Commission (2013) and Ali et al., (2014) had indicated that regional innovation clusters in agribusiness markets have also come up and are composed of large companies and SMEs, start-up companies, public and private research and development centers, universities, specialized suppliers, investors, and regional and government agencies within a geographic region. They work together in a partnership to follow a common nanotechnology in agribusiness markets as regional development strategy, devised to create synergies in specific Nanotech application area (Ortega, 2013). Within the fast-moving agribusiness industry where market competition is very high, continuous development is crucial for the survival of a company (Khodos, 2014). Price discrimination in agribusiness markets is a common behavior where companies charge different prices to different customers as argued by Raynolds et al., (2014). For instance Alishahi et al., (2014) notes that agribusiness companies located within a common region tend to charge varying prices to buyers located in different locales. On this note, Yusuf (2014) also argues that those agribusiness based manufacturers are able to skim the market by product innovation, design and development. Companies in agribusiness markets must continually improve the features, benefits, and performance of their products because of continuous advances in technology, competition, and the changing preferences and needs of customers (Kafi and Ghomi, 2014).

Competitive Analysis of Nanotechnology in Agribusiness Sector

According to Dlamini et al., (2014) agribusiness sector has the potential to attract technological application of nanotechnology as well as herbal medicine industry, both in terms of availability of raw materials, human resources, value addition and economic impact of application of nanotechnology. However, Sekhon (2014) notes that the readiness of food industry will be an internal constraint, owing to the fact that the application of nanotechnology will slightly disrupt majority of existing products which had long been accepted by consumers. The introduction of new products in the market requires a specific cute strategy that is not always easy to create and pursue let alone implementing (Brenes et al., 2014). Although the market in food industry is very large, however in relation to technological application of nanotechnology not all food products can adopt nanotechnology because food is generally served after undergoing thorough cooking process, thus affecting or even damaging the quality of the function of nano itself (Chaudhry et al., 2010; Cota-Arriola et al., 2013; Khodos, 2014). On the other hand, although fertilizer industry in the agribusiness sector has viable natural resources and the level of easiness to facilitate implementation, its readiness, market scale, scope, availability of research and development facilities and human capital still remain undersupplied (Khodos, 2014). This according to Ali et al., (2014) is also contributed to by the ever dependence on the global dynamics, where the price of raw materials and import of energy is still volatile and uncertain. Different forms of government support to the fertilizer industry (in the presence of fertilizer and agricultural subsidies) can raise the potential for competitiveness, although until now, the impact of application of nanotechnology in the fertilizer industry is still not widely reported (Yusuf, 2014). Packaging and pesticides industries have the lowest level of competitiveness because almost all criteria (aspects of technology, raw materials, the readiness of human resources, infrastructure and economic impact and influence of global dynamics, etc.) have relatively no significant value in sustaining the competitiveness of the whole sector. Hirsch et al., (2014) report that competitive analysis has established a very important process that key players in agribusiness sectors have adopted to action their targets. Specifically, individual countries have established different learning centres to foster quality, thus creating an enabling environment for competition and strategic positioning (Wilson et al.,

2014). There are learning plot activities with target groups to explore technological options and to act as "entry points" for learning communities; establishment of "learning and information centres" to capture and explain the experiences of learning groups as well as promotion of farmer-to-farmer training and training of trainers (Rhim et al., 2013; Hirsc et al., 2014). These activities are all geared towards creating stronger attraction into the sector by players with massive capital to invest. In this regard, Brzozowska (2014) asserts that competitive sustainability has been nurtured from diverse position of strengthening of input supply chains. This has seen an improvement in transportation, storage, and distribution; business training for suppliers; involvement of dealers in learning and information centers; formation of input associations, etc. While this may be considered as a weak approach to attain competitive position, experts argue that this model is highly reliable as it will ensure that chain processes are closely interlinked for service delivery, product availability (Kushwaha and Malik, 2012. According ding to Bairwa, Kushwaha, Meena, Lakra, and Kumar (2014) have contended that different players particularly those from European markets have embarked on the process of strengthening their value chains. This has taken shape in the processing, storage, transportation and marketing of targeted products to key consumers as well as to strategic market segments (Bairwa et al., 2014). Encouraging buyer-seller contracts, codes of conduct and industry standard business practices; Hygiene, packaging and labeling practices; strengthening commercial producer associations, are some of the latest competitive processes that have been initiated by the agribusiness sector.

While not forgetting the need to maintain strong financial base, players in food and agribusiness industry have developed models geared towards attracting financial institution into the sector. This has been realized through facilitation of access to credit for farmers and other entrepreneurs involved in the agribusiness cluster through stimulation of savings, guarantee or inventory credit schemes as well as interlinked contracts with buyers. Competitive pressure has also lead to the strengthening of business development services among key agribusiness players and stakeholders as argued by Schiefer and Hartmann (2013). This has been witnessed through capacity building in technical issues (e.g., analysis of commodity value chains, inventory credit systems, market analysis, business planning, etc) (Khodos, 2014). Proper management and decision-making frameworks are also vital tools that foster competitive positions in different markets. In this regard, agribusiness players have developed Managerial issue platforms such as, facilitation and leadership skills, networking and lobbying with the sole intent of maintaining their own competitive position as an industry (Khodos, 2014; Okechukwu, 2014).

DISCUSSION

Based on, the findings of industry trends and dynamics and literature review arguments, it remains relative that Nanotechs technologies and practitioners should adopt different strategic practices in their market strategies so as to become strategically positioned. This will also create a sustained competitiveness within the industry. Reports from Cefic (2006) indicate that Nanotechs industry and market will be influenced by many kinds of societal and economic factors, for instance, the ongoing globalization and population ageing, or changing the workforce and consumer demand. Both these factors lie outside the Nanotechs industry, but with crucial impact. The dynamic of the market of the optics and Nanotechs indicate that the industry is under the influence of several heavy trends (Brayner, 2008). Like all key markets with technological inclination, these heavy backed trends offer the vital driving foundations to describe the optics and Nanotech market growth. The growth of the market of the Nanotechs and the optical being is largely understandable by the rate of adoption of the buyers in search of technology solutions. Nanotechnology is highly regarded by key stakeholders to be a promising area with high impact technological growth and one that has the potential to generate sustainable societal and economic gains. With massive opportunities for gain at stake, competition on the global scene has created a ripple effect among economies and companies to grow and strengthen value of products of nanotechnology. Nanotechnology, as one of the key drivers of the next industrial age, will be critical to the future competitiveness of world's diverse manufacturing sectors, from aerospace to advanced materials to fuel cells to pharmaceuticals and biotechnology (Burns, 2005). Nanotechnology will offer waves of development over time as new discoveries and innovations take place, like how information technology and biotechnology have played out in recent decades. To miss the opportunity to grow in nanotechnology as this field emerges would be a strategic mistake (European NanoBusiness Association, 2005).

Today, vast majority of companies needing to integrate nanotechnologies as they advance are acting as spectators rather than participants. Partly this is because nanotechnology is an emerging scientific field. Much of the action in advancing nanotechnology is taking place in the basic research laboratories found in universities (Bonaccorsi and Thoma, 2005). The key question is whether companies will be able to gain the skill sets and knowledge of advancing nanotechnology and whether such accumulation of skills will lead to strong positioning and competitive advantage within Nanotechnology markets (Schwab, 2009). Accordingly, nanotechnology innovations are one of the sources of the competitive advantage of a country (Porter, 1990). In this regard, the identification and analysis of competitiveness at the global level in modern conditions have become necessary. One of the components of macro-research involves the competitiveness of countries, companies and other market participants. "A nation's competitiveness depends on the capacity of its industry to innovate and upgrade" (Porter, 1990), while a country's competitive behavior reflects on its competitive position and status. Competitive status mirrors the competitive behavior of all market participants (Kotler et al., 2007) that are positioned competitively, protecting or enhancing their status by strategies.

Competitive status of a country in nanotechnology is determined by two indicators - the level of its nanotechnology activities (nanotechnology innovation) and the power of its technology development (Burns, 2005). The level of nanotechnological activity points to "the capabilities and resources of a nation's engine for nanotech innovation (Tang and Shapira, 2011), and the power of technological development points to the possibility of a country to develop its economy on nanotechnology (Burns, 2005).

In nutshell, the alternative indicators may not present an accurate view of technological leadership and economic competitiveness for many reasons. Nor does global technological leadership alone guarantee that the economic value produced by nanotechnology innovations will be captured within international markets (Kotler et al., 2007). In today's global economy, companies have the option on of locating work — e.g., research, development, design, engineering, manufacturing, product support — where it can be done most effectively (Kotler et al., 2007). This is the only formidable way to achieve high competitive strategies in nanotech market positioning and competitiveness (Schwab, 2009). The competitive and positioning advantages can be attained through proper study of market trends, market dynamics and the eventual formulation of effective market strategies, as highlighted by the research model in the appendix section. With regards to market trends, technology and market trends significantly affect network architectures, economies of scale and direct systems cost, ultimately influencing the overall future performance and price of nanotechnological products (Shapira and Wang, 2009). The market dynamics also shows that in the fast-moving Nanotech industry where market skimming is possible; price discrimination, new product design and development is crucial for the survival of a company. Strategic positioning for competitive advantage in nanotechnology markets also requires a close look at the strategic process (Tang and Shapira, 2011). Nanotech technologies and practitioners in different nations should adopt different strategic practices in their market strategies so as to become strategically positioned (Porter, 1990). This will also create a sustained competitiveness within the industry (Tang and Shapira, 2011). Nanotech is a very dynamic and vibrant industrial sector that holds the potential for huge market growth. The expected compound annual growth rate for Nanotechs over the coming years is 8%, clearly demonstrating the rapid growth of this key technology sector. Major growth is expected in lighting, medical technologies and life sciences, laser-based manufacturing and optical communications (Brenes et al., 2014). The growth of the global Nanotechs industry is expected to accelerate the trends. The recent dramatic expansion of the optical telecommunications markets is only one aspect of this. Other emerging areas such as Bio-Nanotechs and Nanotechnology are likely to have an equally large impact over the next five to ten years. It is critical for companies to identify these trends and ensure that their products are properly aligned to benefit from these developments, rather than become crushed by them. Technology and market trends significantly affect network architectures, economies of scale and direct systems cost, ultimately influencing the overall future performance and price of nanotech networks.

PRACTICAL IMPLICATIONS, SOCIAL AND ECONOMIC CONTRIBUTIONS

Interests and motivation towards nanotechnology has significantly grown in the last 7 to 8 years, and this scientific field has proved to be particularly promising in the development of an emerging, disruptive and rapidly-growing market for nanotech applications in agribusiness sector. This chapter offers a picture of the overall global market for nanotech applications in agribusiness sector and provides an insight into the management and organization of key stakeholders operating in such a market. The propositions of this chapter have significant implications to both academic scholars and practitioners: First, it is believed that the present research can benefit researchers interested in the study of the nanotech market and, in general, the high-tech industry applications in agribusiness sector.

Second, the chapter suits as a background policy document for policy makers and investors of nanotech applications in agribusiness sector. The high-tech disruptive industries are increasingly paying close attention to the design of public policies (U.S. Government, 2006, 2003), and particularly the nanotech industrial application in agribusiness sector is considered critical in favoring the development of advanced economies (European Commission, 2004c). In this respect, the chapter's achievements are useful since they suggest how to build a system of supporting initiatives that fits with the current evolution and characteristics of the world nanotech industry. These initiatives, in fact, are likely to be far more effective, within the specific stage of the industry's life cycle, if they aim at stimulating the diffusion of those firms that adopt the types of disruptive business models which are contextually capable of generating and delivering the highest value addition.

Third, the chapter provides corporate executives with useful insights into the role played by nanotechnology within agribusiness context and the way nanotech companies are currently managing and organizing their business activities to favour this industry. In order to define the optimal business model for their companies, the executives of firms aimed at commercializing nanotech applications in agribusiness sector should first consider the current attributes of nanotech market and its stage of development, and then adapt the role they seek to play in the agribusiness market to a figure that is likely to be precursory of the highest value.

Future Projection of Nanotechnology in Agribusiness Sector

The dynamic nature of agribusiness sector provides significant future business challenges and opportunities for future watch. The expected growing demand for food by itself presents potential sales and revenue growth in the agribusiness sector. Additionally, the expected future development of expanding bio-economy with biological based raw materials being used in the energy, industrial and health/pharmaceutical industries adds further potential (Mukhophadhyay, 2014; Brenes et al., 2014; Ranjan et al., 2014). According to Hirsch et al., (2014) the integration of the agricultural sector into the broader overall global industrial economy creates opportunities for new innovative product and service offerings as well as new value chains and markets to deliver those new products and services (Raynolds, et al., 2014). It adds further complexity to an already complex value chain. But according to Dlamini et al., (2014), that future also is highly uncertain. Many of the management implications of the challenges of strategic uncertainty, innovation and structural change have been identified earlier; however, they can further be highlighted (Arest et al., 2013). Strategic uncertainty requires managers to develop additional capacities to monitor the business climate in which they operate, to anticipate as best one can, the impact of the highly improbable, and to regularly reassess the firm's strategic positioning to capture unexpected opportunities and mitigate potential catastrophic losses (Plascencia-Jatomea, 2013; Ranjan et al., 2014; Ali et al., 2014). This may require a more flexible rather than focused strategy and a real options mentality embracing more experimentation rather than making "full-blown" or "big bet" commitments (Sastry et al., 2013; Rhim et al., 2013). As to innovations, searching out potentially disruptive technologies or innovations and assessing the risk and rewards of being a first mover vs. fast follower in the commercialization of those technologies or innovations will be critical to capture market potential or defend against new entrants (Brenes et al., 2014). Systematic and frequent stage-gate processes to evaluate the success potential of innovations as they move from a new idea or invention to commercialization will reduce the risk and enhance the probability of success from innovation (Magnuson et al., 2011; Sodan and Vernaeau, 2014). Ali et al., (2014) note that criteria such as potential return, market un-certainty, technical/regulatory uncertainty, time to market, access to capabilities, and costs al-ready incurred should be included in the selection methods used by companies. On the same note Kafi and Ghomi (2014) argue that food and agri-business companies should also rely on several selection methods and on an assessment of the projects by cross-functional teams as well. Finally, systematically documenting the knowledge created in the innovation process will increase the value created irrespective of whether the product/service offering is a commercial success – learning from and communication on an un-successful innovation or venture has the potential to improve the chances of success in future innovations/ventures (Khodos, 2014; Yusuf, 2014). Finally, the significant structural changes in the agribusiness sector suggest that managers need to be increasingly vigilant in assessing the competition they will face as well as the opportunities they may have in shaping the restructuring of their industry (Alishahi et al., 2014). The evolution of new value chain structures and industry convergence will require additional leadership and management skills along with new relationships and linkages outside of what have been historical industry boundaries (Sekhon, 2014).

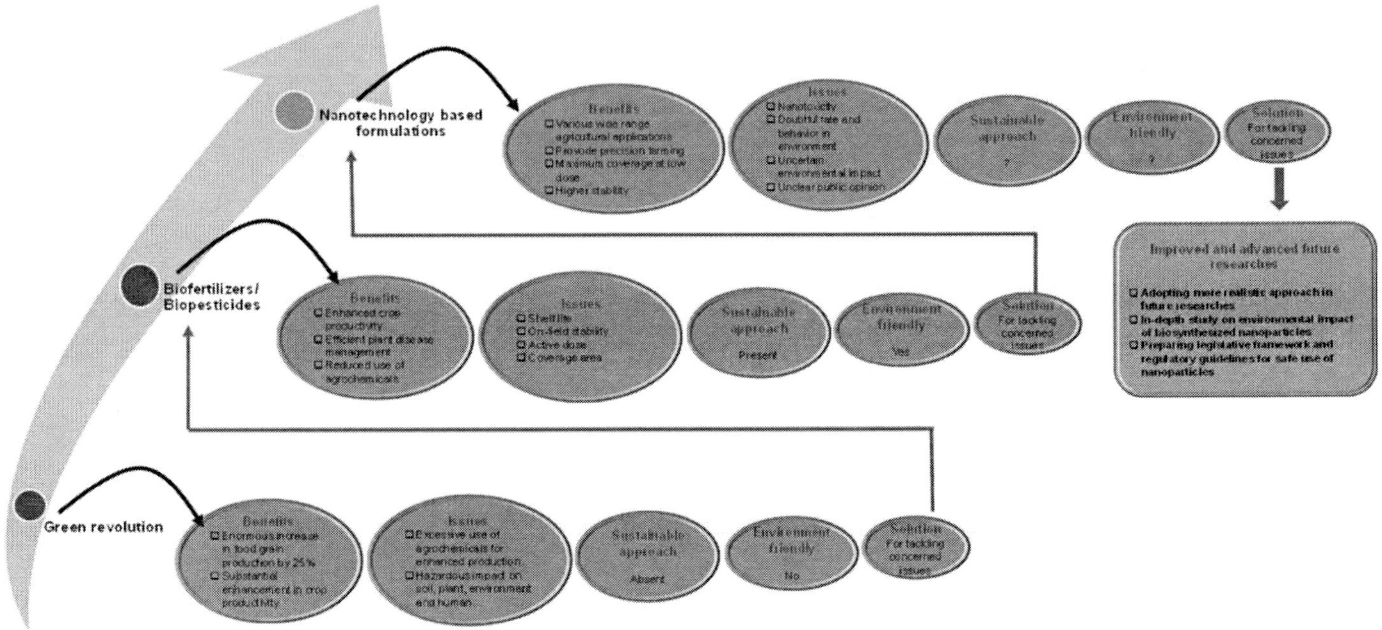

Source: Mishra, Sandhya & Keswani, Chetan & C. Abhilash, P & Fraceto, Leonardo & Singh, Harikesh. (2017). Integrated Approach of Agri-nanotechnology: Challenges and Future Trends. *Frontiers in Plant Science*. 8. 10.3389/fpls.2017.00471.

Figure 4. Recommendations on adoption of nanotechnology in agribusiness sector.

The information, knowledge base and skill set for analyzing and understanding these issues, and making the critical strategic decisions to be successful in an increasingly turbulent business climate, requires integration of concepts from economies, management, finance, decision sciences, organizational behavior, and strategy, as highlighted in figure 4 below. Our goal here has been to make a modest contribution to that knowledge base with a focus on strategic uncertainty, innovation and structural change in the agribusiness sector.

CONCLUSION

Strategic thinking will become increasingly important in the successful activation of agribusinesses sector in the future with respect to nano-technology applications in agriculture (Brenes, Montoya and Ciravegna, 2014). Strategic thinking process leverages on an understanding of the strengths and weaknesses of a business – leading to an identification of the core competencies and capabilities – to respond to opportunities and threats provided by the external environment (Bukucs et al., 2014). Only by understanding internal capacities and the external environment, can the successful agribusiness manager analyze the strategic positioning options available to the business of agricultural outputs. Choosing a strategic position of operational excellence/low-cost leader, product/process innovation or customer intimacy based on this analysis provides the manager with a fundamental direction for the business that can shape its ability to create value for its customers. And that is what the successful agribusiness of tomorrow will be a business that understands customers and determines how to create value for those customers through products and services that create a long-term sustainable competitive advantage through an operational excellence/low cost, product/process innovation or customer intimacy strategy.

References

Ali, M. A., Rehman, I., Iqbal, A., ud Din, S., Rao, A. Q., Latif, A. and Husnain, T. (2014). Nanotechnology: A new frontier in Agriculture. *Advancements in Life Sciences,* 1(3), 129-138.

Alishahi, A., Proulx, J. and Aider, M. 2014. Chitosan as Bio-based Nanocomposite in Seafood Industry and Aquaculture. *Seafood Science: Advances in Chemistry, Technology and Applications,* 211.

Amna, T. and Hassan, M. S. (2013). Inactivation of Foodborne Pathogens by NiO/TiO2 Composite Nanofibers: A Novel Biomaterial System. *Food and Bioprocess Technology,* 6, 988-996.

Aresta, A., Calvano, C. D., Trapani, A., Cellamare, S., Zambonin, C. G. and De Giglio, E. (2013). Development and analytical characterization of vitamin(s)-loaded chitosan nanoparticles for potential food packaging applications. *Journal of Nanoparticle Research*, 15:1592-1597.

Bairwa, S. L., Kalia, A., Meena, L. K., Lakra, K. and Kushwaha, S. (2014). Agribusiness Management Education: A Review on Employment Opportunities. *International Journal of Scientific and Research Publications (IJSRP),* 4(2), 1-4.

Bairwa, S. L., Kushwaha, S., Meena, L. K., Lakra, K. and Kumar P. (2014). Agribusiness Potential of North Eastern States: A SWOT Analysis. Edited by Singh et al. *Agribusiness Potentials in India: experience from hill states.* EBH Publishers (India) Guwahati New Delhi. PP 544–556.

Bakucs, Z., Fałkowski, J. and Fertő, I. (2014). Does Market Structure Influence Price Transmission in the Agro- food Sector? A Meta-analysis Perspective. *Journal of Agricultural Economics,* 65(1), 1-25.

Banterle, A., Cavaliere, A., Carraresi, L. and Stranieri, S. (2014). Food SMEs Face Increasing Competition in the EU Market: Marketing Management Capability Is a Tool for Becoming a Price Maker. *Agribusiness,* 30(2), 113-131.

Bonaccorsi, A. and Thoma, G. (2005). *Scientific and Technological Regimes in Nanotechnology: Combinatorial Investors and*

Performance, Laboratory of Economics and Management, Sant'Anna School of Advanced Studies, Pisa.

Brayner, R. (2008). "The toxicological impact of Nanoparticles." *Nanotoday.* Vol. 3, No. 1-2.48-55.

Brenes, E. R., Montoya, D. and Ciravegna, L. (2014). Differentiation strategies in emerging markets: The case of Latin American agribusinesses. *Journal of Business Research,* 67(5), 847-855.

Brzozowska, A. (2014). *Management, Budgeting and Evaluation of the Rural Development Programme in Agribusiness Logistic Systems in Poland.*

Bucheli, T. D., Knauer, K., Gogos, A. (2013). Overview of agricultural applications of nanotechnology; Nanotechnology for the agricultural sector: from research to the field; *Workshop organized by JRC IPTS Agrilife Unit,* 21 – 22 November 2013, Seville, ES.

Burns, R. (Ed.). (2005). Ranking the Nations: Nanotech's Shifting Global Leaders, Statement of Findings: Ranking the Nations: Nanotech's Shifting Global Leaders. *Lux Research.* Retrieved from http://israelsciencetechnology.blogspirit.com/files/lux_research_sof_nts-r-05-006.pdf.

Chandrasekaran N. (2013). Distinctive effects of nano-sized permethrin in the environment. *Environmental Science and Pollution Research,* 20, 2593-2602.

Chen, L., Zheng L., Lv, Y., Liu, H., Wang, G., Ren, N., Liu, D., Wang, J. and Boughton, R. I. (2010). Chemical assembly of silver nanoparticles on stainless steel for antimicrobial applications. *Surface Coatings Technol.*, 204, 3871.

Chin, C. P., Wu, H. S. and Wang, S. S. (2011). New Approach to Pesticide Delivery Using Nanosuspensions: Research and Applications. *Industrial and Engineering Chemistry Research,* 50, 7636-7643.

Choudhury, S. R., Nair, K. K., Kumar, R., Gogoi R., Srivastava, C., Gopal, M., Subhramanyam, B. S., Devakumar, C. and Goswami, A. (2010). Nanosulfur: A potent fungicide against food pathogen, *Aspergillus niger. AIP Conference Proceedings,* 1276, 154-157.

Cota-Arriola O., Onofre Cortez-Rocha, M., Burgos-Hernández A., Marina Ezquerra-Brauer, J. and Plascencia-Jatomea, M. (2013). Controlled release matrices and micro/nanoparticles of chitosan with antimicrobial potential: Development of new strategies for microbial control in agriculture. *Journal of the Science of Food and Agriculture,* 93, 1525-1536.

Danisch, E. P. A. (2013). Systemic Absorption of Nanomaterials by Oral Exposure. *Report Environmental Project* No. 1505, 2013.

Das, S. K., Khan, M. R., Parandhaman, T., Laffir, F., Guha, A. K., Sekarana, G. and Mandal, A. B., (2013). Nano-silica fabricated with silver nanoparticles: antifouling adsorbent for efficient dye removal, effective water disinfection and biofouling control. *Nanoscale,* 5, 5549-5560.

De Azeredo, M. C. H. (2013). Antimicrobial nanostructures in food packaging, *Trends in Food Science and Technology*, 30, 56–69.

De Stefano, D., Carnuccio, R. and Maiuri, M. C. (2012). Nanomaterials Toxicity and cell Deatch Modalities. *J. Drug Discovery,* doi: 10.1155/2012/167896.

Debnath, N., Das, S. and Goswami A. (2011). Novel entomotoxic nanocides for agro-chemical industry. *Proceedings of the IEEE Conference on Nanotechnology,* pp 53-56.

Dlamini, B. P., Kirsten, J. F. and Masuku, M. B. (2014). Factors Affecting the Competitiveness of the Agribusiness Sector in Swaziland. *Journal of Agricultural Studies,* 2(1), 61-72.

Duraccio D., Silvestre C., Pezzuto M., Cimmino, S. and Marra, A., (2013). Polypropylene and Polyethylene-Based Nanocomposites for Food Packaging Applications. In: *Ecosustainable Polymer Nanomaterials for Food Packaging,* CRC Press, pp. 143-168.

Durán, N. and Marcato, P. D. (2013). Nanobiotechnology perspectives. Role of nanotechnology in the food industry: a review. *International Journal of Food Science and Technology,* 48(6), 1127-1134.

EFSA Scientific Committee. (2011). Guidance on the risk assessment of the application of Nanoscience and nanotechnologies in the food and feed chain. *EFSA Journal* 9 (5): 2140.

European Commission (DG SANCO). (2013). Assessment of nanomaterials in food, health and consumer products. *Workshop on the Second Regulatory Review on Nanomaterials*. D. H. a. Consumers. Brussels.

European Food Safety Authority, (EFSA). (2012). *Annual report of the EFSA Scientific Network of Risk Assessment of Nanotechnologies in Food and Feed for 2012.* Parma, Italy: European Food Safety Authority.

Gogos A., Knauer K. and Buchell, T. D. (2012). Nanomaterials in plant protection and fertilization: Current state, foreseen applications, and research priorities. *Journal of Agricultural and Food Chemistry*, 60, 9781-9792.

Hirsch, S., Schiefer, J., Gschwandtner, A. and Hartmann, M. (2014). The Determinants of Firm Profitability Differences in EU Food Processing. *Journal of Agricultural Economics.*

Hutchison, J., E., (2008). Greener Nanosciences: a proactive approach to advancing applications and reducing implications of nanotechnology, *ACS Nano.* 2: 395-402.

Jałowiecki, P., Woźniakowski, T. and Ząbkowski, T. (2014). *Some Remarks on Logistics Investments Among Polish Food Processing and Agribusiness Companies.* Beata Butryn, Piotr Machura, 122.

Kafi, F. and Ghomi, F. S. (2014). *A Game-Theoretic Model to Analyze Value Creation with Simultaneous Cooperation and Competition of Supply Chain Partners*. Mathematical Problems in Engineering, 2014.

Khodos, D. (2014). Economic Mechanism of Innovative development of Agricultural Sector. *Innovation Management and Company Sustainability,* 26, 73.

Kole C., Kole P., Manoj Randunu K., Choudhary P., Podila K., Chun Ke P., Rao, A. M. and Marcus, R. K. (2013). Nanobiotechnology can boost crop production and quality: first evidence from increased plant biomass, fruit yield and phytomedicine content in bitter melon (*Momordica charantia*). *BMC Biotechnol,* 13.

Kotler, P. H., Vong, V., Sonders, D. Z. and Armstrong, G. 2007. *Principles of Marketing.* Zagreb. Mate.

Kuan, C., Y. and Yee-Fung W. (2012). Nanotech: Propensity in foods and bioactives. *Critical Reviews in Food Science and Nutrition,* 52, 55-71.

Kushwaha, H. B. and Malik, C. P. (2012). Contributions of Nanotechnology in agriculture and food processing. *Phytomorphology: An International Journal of Plant Morphology,* 62, 57-67.

López-Salazar, A., López-Mateo, C. and Molina-Sánchez, R. (2014). What Determines the Technological Capabilities of the Agribusiness Sector in Mexico? *International Business Research,* 7(10), p 47.

Lu, J. and Bowles, M. (2013). How Will Nanotechnology Affect Agricultural Supply Chains? *International Food and Agribusiness Management Review Volume 16*, Issue 2, 2013.

Magnuson, B. A., Jonaitis, T. S. and Jeffrey, W. (2011). A Brief Review of the Occurrence, Use, and Safety of Food-Related Nanomaterials. *Journal of Food Science.* 76 (6):R126-R133.

Mukhopadhyay, S. S. (2014). Nanotechnology in agriculture: prospects and constraints. *Nanotechnology, science and applications,* 7, 63.

Narayanan A., Sharma P. and Moudgil, B. M. (2012). Applications of engineered particulate systems in agriculture and food industry. *KONA Powder and Particle Journal*, 30, 221-235.

Okechukwu, O. T. (2014). Critical Factors Influencing the Entrepreneurial Undergraduate's Decision Venturing into Agribusiness in Ebonyi State, (A Case Study of Ebonyi State University). *Journal of Natural Sciences Research, 4*(19), 33-38.

Ortega, D. L. (2013). Modernizing Agrifood Chains in China: Implications for Rural Development. *American Journal of Agricultural Economics,* 95(4), 1046-1048.

Papaspyrides, C. C. D. (2010). Nanotechnology and food contact materials, *AIP Conference Proceedings,* 1255, 234-236.

Porter, M. (1990). The Competitive Advantage of Nations. *Harvard Business Review*, 68 (2), 73-93.

Rama, R. and Martínez, C. (2013). The changing structure of the global agribusiness sector. *Handbook of Industry Studies and Economic Geography,* 305.

Ranjan, S., Dasgupta, N., Chakraborty, A. R., Samuel, S. M., Ramalingam, C., Shanker, R., Kumar, A. (2014). Nanoscience and nanotechnologies in food industries: opportunities and research trends. *Journal of Nanoparticle Research:* 16: 2464-2487.

Rashidi, L., Khosravi-Darani, K., (2011). The Applications of Nano-technology in Food Industry. *Critical Reviews in Food Science and Nutrition:* 51:723-730.

Raynolds, L. T., Long, M. A. and Murray, D. L. (2014). Regulating Corporate Responsibility in the American Market: A Comparative Analysis of Voluntary Certifications. *Competition and Change,* 18(2), 91-110.

Reardon, T. C., Barrett, Berdegué J. and Swinnen, J. (2009). Agrifood Industry Transformation and Small Farmers in Developing Countries. *World Development;* 37 (11): 1717-1727.

Rhim, J. W., Park, H. M., Ha, C. S. (2013). Bio-nanocomposites for food packaging applications. *Progress in Polymer Science:* 38(10):1629-1652.

Sastry, R. K., Anshul, S., Rao, N. H. (2013). Nanotechnology in food processing sector- An assessment of emerging trends, *Journal of Food Science and Technology* 50: 831-841.

Schiefer, J. and Hartmann, M. (2013). *Industry, firm, year, and country effects on profitability in EU food processing.* Institute for Food and Resource Economics, University of Bonn. Discussion Paper, 2.

Schwab, K. (Ed.). (2009). The Global Competitiveness Report 2009-2010. Switzerland, Geneva. *World Economic Forum.* Retrieved from https://members.weforum.org/pdf/GCR09/GCR20092010fullreport.pdf.

Scoones, I., Smalley, R., Hall, R. and Tsikata, D. (2014). *Narratives of scarcity: understanding the 'global resource grab'.*

Sekhon, B. S. (2014). Nanotechnology in agri-food production: an overview. Nanotechnology, *Science and Applications* 7: 31-53.

Shapira, P. H. and Wang, J. (2009). From lab to market? Strategies and issues in the commercialization of nanotechnology in China. *Asian Business and Management,* 8 (4), 461–489. doi: 10.1057/abm.2009.15.

Silvestre, C., Cimmino, S. (2013). *Ecosustainable Polymer Nanomaterials for Food Packaging*, CRC Press, New York, US.

Sodano, V. and Verneau, F. (2014). Competition Policy and Food Sector in the European Union. *Journal of International Food and Agribusiness Marketing*, 26(3), 155-172.

Srivastava, S., Mathur, D. H. and Kumar, M. (2014). Competitive Advantage and Value Chain Analysis of Small Scale Fruits and Vegetable Processing Industry. *International Journal of Logistics and Supply Chain Management Perspectives*, 2(4), 778-790.

Tang, L. and Shapira, P. H. (2011). Regional development and interregional collaboration in the growth of nanotechnology research in China. *Scientometrics* 86 (2), 299-315. doi 10.1007/s11192-010-0274-9.

Tilman, D., Balzer C., Hill, J., Befort, B. L. (2011). Global food demand and the sustainable intensification of agriculture, *Proceedings of the National Academy of Sciences*, 108: 20260–20264.

Timmer, P. (2010). *Lessons from Food Crises Past.* Center for Global Development, Washington D.C.

Tyler, L. C. (2012). "FDA Issues Draft Guidance on Use of Nanotechnology in Food and Food Packaging." *Nanotechnology Law and Business* 9: 149-155.

Vink, N. (2010). Agribusiness Policy and Strategy in the Future. *2010 Conference of the Agricultural Business Chamber (ABC),* Lord Charles Hotel. Somerset West, South Africa.

Wilson, W. W., Shakya, S. and Dahl, B., (2014). *Dynamic Changes in Spatial Competition for Fertilizer.*

World Bank. (2013). *Growing Africa: Unlocking the Potential of Agribusiness.* AFTFP/AFTAI. World Bank: Washington D.C.

Yusuf, S. A. (2014). *The Analysis of Export Performance of Newly Industrialized Countries (NICs):* The Lesson for African Countries.

In: Agribusiness
Editor: Ronald Luna
ISBN: 978-1-53614-413-0

Chapter 3

MEXICAN COFFEE AGRIBUSINESS, TRENDS, CHALLENGES AND GOALS TOWARDS SUSTAINABILITY

Tania Marín-Garza[1], Noe Aguilar-Rivera[1,*] and Odón Castañeda-Castro[2]
[1]Facultad de Ciencias Biológicas y Agropecuarias,
Camino Peñuela-Amatlán de los Reyes S/N, Veracruz, México
[2]Universidad Veracruzana, Facultad de Ciencias Químicas,
Prolongación de Oriente 6 No. 1009. Orizaba, Veracruz, México

ABSTRACT

Coffee is considered the economic engine for many of the producing countries, this way, coffee is the second most traded product in the world after oil, it is the most consumed beverage in the world, after water. The total surface of green coffee harvested in the world is more than 10 million hectares and 145 million bags, out of which 7% was originated in Mexico, which makes the country occupy the fifth place with 3.5% of the world coffee supply and thus has great productive roots and high consumption. Coffee-growing is carried out in 675 thousand hectares,

* Corresponding Author Email: naguilar@uv.mx.

with yields 382 kg ha-1 and which are concentrated in the areas of the Gulf Slopes, the Pacific Ocean Slopes, the Soconusco Region as well as the North Central Region of Chiapas and in other 12 states of the country, out of which 74% of the overall production is generated in Chiapas, Oaxaca and Veracruz, along with 404 municipalities, 4,572 localities in altitudes ranging from 300 to 2000 masl, but with indicated elevations of 600 to 1200 masl, and quite higher than other 510 producers. The production of coffee as agribusiness for the country is an important source of foreign exchange and income for thousands of small producers and around 30 indigenous groups, during the current harvest between the years of 2018 and 2019, there is an estimated production of 4.5 million kilos of coffee. Currently, a sustainable crop has been promoted that allows the use of coffee residues, creating added value as fertilizer, biofuel, substrate for various crops, and more recently, in the extraction of caffeine and polyphenols.

The objective of this chapter is to assess and measure the current growth on the agribusiness of coffee in Mexico, its trends, its challenges, the technology being used, and of course, the constraints and goals towards the sustainability of the value chain and the utilization and application of the main coffee industry wastes (coffee pulp and husk) and the possibility of utilization of such, as raw materials for bio-refinery (energy, chemical and food) are discussed hereafter.

1. Introduction

Coffee growing in Mexico is considered a fundamental strategic activity, due to the fact that it allows the integration of productive chains, generation of foreign exchange and employment, the way of subsistence of many small producers and over 30 indigenous groups that in recent times have had ecological relevance, since more than 90% of the area planted with coffee is under diversified shades. However, the relevance of the coffee sector has been also immersed in the recurrent crises due to the substantial drop of prices in the international market, which is why options are sought and strategies are focused on the production of organic and fair trade of coffee, with the objective to develop more competitive advantages, this production will have greater potential as it triggers or registers the broader processes of rural development (Escamilla-Prado et al., 2015, 357-358).

Coffee production in Mexico reports decreasing trends. In the current decade, the national harvest was reduced to an annual average rate of 4.2%, which reached its lowest level since 1980 (see Figure 1). The main factor that explains this type of decrease in national production is the sustained reduction in the productivity of coffee plantations, as well as the effects caused by pests of great importance such as coffee rust. On the other hand, coffee consumption in Mexico continues to rise, favored by the increase in the number of specialized establishments and other points of sale of this beverage (FIRA, 2016).

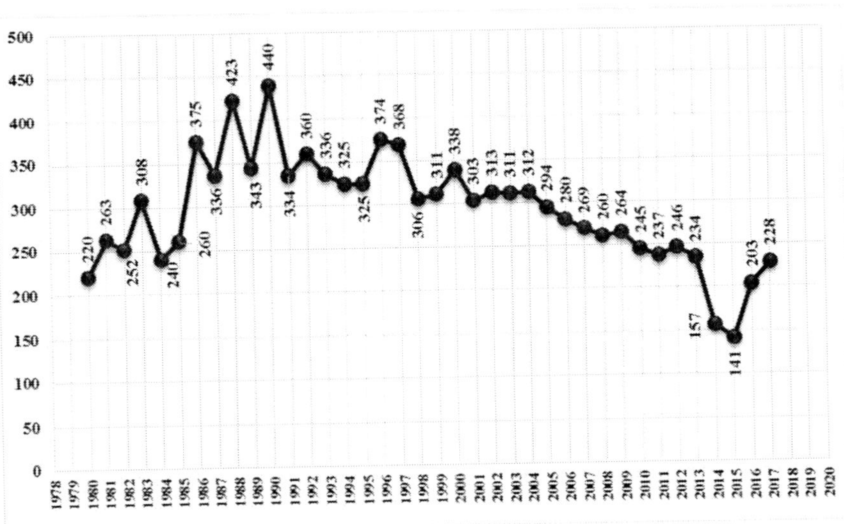

Source: Provided by SAGARPA, 2017, FIRA, 2016.

Figure 1. Production of green coffee in thousands of tons

1.1. Background

Coffee is the second most traded product in the world, after oil, which is why it is considered the economic engine for many countries, it is currently the most consumed beverage right after water (Oliveira et al., 2011, 330-337), which allows observing two things, first that many people depend economically on this type of beverage and second, the large number of people who are addicted to it (National Federation of Coffee Growers of Colombia, 2018, CEDRSSA, 2016).

1.2. Origins of Coffee

The coffee tree is native to Africa, its name is derived from the city of Kaffa, although the history of its origin is not well stablished, there is an almost unanimous consensus that it originated in the wild regions, among the highlands of Abyssinia, now Ethiopia (ICO, 2017), the evidence that constitutes this hypothesis is that Arabica coffee grows wild in the neighboring areas of Sudan, at altitudes higher than 1,500 meters above sea level (León, 2000, 350-364).

Although part of the history of coffee is considered a legend, what is known with greater certainty is that the slaves that were taken, from the current Sudan to Yemen and from there to Arabia through the large port in Moca, it is now well known as synonymous of coffee, they would ingest the succulent meaty part of the coffee cherry, so it is undoubted that coffee was grown in Yemen in the fifteenth century and from there it was introduced in other regions such as Arabia and Egypt, where it began to acquire fame and value, for several centuries the crop remained a monopoly of the Arabs in the vicinity of the Red Sea (ICO, 2017, González de Mejía and Ramírez-Mares, 2014, 2-4).

1.3. Botany and Physiology

The common name given to the coffee plant is coffee and it belongs to the genus of trees of the *Rubiaceae* family, which covers 500 genera and 8,000 species (Contreras and Hernández-Martinez, 2008, 9-10). One of these genera is the *Coffea*, which consists of trees, shrubs, and vines, all native to tropical Africa and some islands of the Indian Ocean, such as Madagascar (Waller et al., 2007, 327.328) and comprises about 10 civilized species, that is, cultivated by men, and 50 wild species

(CENICAFE, 2018), it's a perennial shrub which life cycle in commercial conditions reaches between 20 to 25 years, depending mainly on agronomic management, the pyramidal plant begins to produce fruit in the buds of the knots of the primary branches that give rise initially to the leaves in the shape of a number of two and in opposite form. Subsequently, the flowers emerge from the fruiting buds, located at the insertion point of the leaves on the branches, which then become fruit, after a year and continues its production for several years until reaching its maximum productivity, between six and seven years of age (MIDA, 2015).

Taxonomically, the coffee belongs to the Plantae kingdom, *Espermatophytes* type, Sub-type Angiosperms, Dicotyledonous class, Subclass *Giopetrata inferiovariadas*, *Rubiales* order, *Rubiaceae* family, genus *Coffea*, *Eucoffea* subgenus, *arabica* species, Canephora, Liberica, Deweri (ICO, 2017).

The coffee species which possess an economic importance, botanical properties and cup quality, based on the demands of the international market, are: *Coffea arábica* (*C. arabíca* L.), *Coffea liberica* (*C. liberica* Mull ex Hiern), *Coffea canéfora* (*C. canephora* Pierre ex Froehner) and *Coffea dewevrei* or coffee excelsa (*C. excelsa* A. Chev.). 85% of coffee production in the world cultivates *Coffea arabica* and the remaining 15% grows species or varieties of economic importance for each country (FUNDESYRAM, 2010, INFOAGRO, 2017, ICO, 2017, COFFEEBI, 2018).

1.3.1. Coffea arabica

Arabica coffee, (*C. arabica* L., syn: *C. vulgaris* Moench, *C. laurijolia* Salisb.) is native to the highlands of Ethiopia, at elevations ranging between 1,350 and 2,000 meters above sea level. It is possibly native to other parts of Africa and Arabia; it was first described in 1753 by Linnaeus. It is genetically different from other coffee species, since it is tetraploid, which makes it have a total of 44 chromosomes instead of 22. It is a shrub about five meters high, with leaves relatively small, but vary in width size, averaging 12 to 15 cm long and about six cm wide, typically oval or elliptical, acuminate, short, sharp at the base, sometimes somewhat undulating, flowering takes place right after the rainy season and its flowers are arranged in a cluster-like shape. The berry is oblong-elliptical, measuring about 1.5 cm in length, green at the beginning, and afterwards, turn red when they mature, after an average of seven to nine months. The seeds vary in size from 8.5 to 12.7 mm in length, Arabica coffee is most often susceptible to pests and diseases, so obtaining resistance is one of the main objectives of plant improvement programs, this species of coffee is grown throughout Latin America, Central and Eastern Africa, India, Indonesia, Brazil and Colombia being the largest producers of this species, which is used as roasted and ground coffee, and it is added in mixtures with Robusta to improve the quality of soluble coffee (Saucedo-García et al., 2014, 1-6; Mehari et al., 2016, 1-16).

1.3.2. Coffea Canephora

Robusta coffee (*C. canephora* Pierre ex Froehner) is native to the equatorial forests of Africa, it grows in elevations from sea level to about 1,000 masl, it is a sturdy smooth tree, with shallow roots that can reach up to 10 meters in height, with broad leaves that sometimes acquire a corrugated or wavy, elliptical, short, acuminate appearance. It has white flowers, sometimes slightly diffuse with a pink color, in two axillary clusters, sessile, with or without bracts with leaves. The fruit has a rounded shape and it takes up to 11 months to mature, its elongated seed is widely ellipsoid, between 8 to 16 mm in size, it is cultivated in Central and Western Africa, throughout Southeast Asia and lesser amount in some regions of Brazil, where it was named Conillón, the price and the quality of the Robusta grain is much more inferior to the Arabica varieties, ranging between 30 to 40%, its main use is in soluble coffee and in mixtures with

Arabica to add body and cream to the cup, Robusta coffee and its hybrids have greater tolerance to blight caused by *Hemileia* Spp., and other diseases, it produces more grain than the Arabica and maintains better mature fruit in the tree, the main producers of Robusta coffee are Vietnam, Brazil and Indonesia (MIDA, 2015).

1.3.3. Coffea Liberica

Liberian coffee (*C. liberica* Bull ex Hiern) is native to the surroundings of Monrovia in Liberia, and its discovery dates from the late nineteenth century, it's a highly ornamental tree and was soon known in many other parts of the world, after its discovery in 1872, it is a strong and large tree, up to 18 meters in height, with large and leathery-like leaves. The fruit and the seed (grain) are twice the size of the Arabica, it resists sun exposure better than most other members of the genus and has good retention of its mature fruits; It has a proportion of fresh fruit in relation to dry grains of about 10:1, it grows well at high altitudes, and it offers a coffee with a peculiar flavor, so its consumption is not very widespread and it is only sold in very small quantities. however, is particularly appreciated in the Scandinavian countries, in the Ivory Coast there are many other varieties of Liberian coffee, among them liberiensis and indeniensis can be mentioned herein, the last one being the one that is most cultivated (Hurtado and Herrera Pinilla, 2013, 32-36).

1.4. Chemical Characteristics of Coffee

The chemical characteristics of coffee vary according to the species, and roasting, in the Arabica variety, caffeine includes 1.2% of the dry matter, 4.2% of minerals, out of which 1.7% is potassium, 16.0% lipids, 1.0% trigonelines, 11.5% proteins and amino acids, 1.4% aliphatic acids, 6.5% chlorogenic acids, 0.2% glycosides and 58.0% carbohydrates. In the Robusta variety, caffeine comprises 2.2% of the dry matter, 4.4% minerals, of which 1.8% correspond to potassium; 10.0% lipids, 0.7% trigonelines, 11.8% proteins and amino acids, 1.4% aliphatic acids, 10.0% chlorogenic acids and 59.5% traces glycosides and carbohydrates. The water content in commercial raw coffee beans varies between 8.0 and 12.0% (Echeverri et al., 2005, 360-361). However, these parameters vary due to factors such as the species, variety, nature and fertility of the soil, the atmospheric and environmental conditions, the agronomic management, the age of the plantation, pests, diseases and their control, as well as the times of harvest, the moisture content of the cherry coffee and the harvest which influence the yields and the characteristics of the grain; Furthermore, an efficiently cultivated coffee can lose its inherent qualities as a consequence of an inadequate process of profit (Díaz and Riaño, 2017, 34-40).

1.5. Cultivation of Coffee

For the cultivation of coffee, there are defined agroecological characteristics, in their different phenological phases (DaMatta et al., 2007), greater possibilities will have to express all their genetic potential, which will result in greater production, cup quality which is what ultimately the coffee farmer is mainly interested in (López-García et al., 2016, 298-303).

1.5.1. Coffea arabica

The best *Arabica* coffee plantations in Mexico are found at altitudes between 900 and 1200 masl, because it is in this range that you can obtain excellent quality coffee, due to its organoleptic characteristics such as taste, aroma, body and acidity. The average annual rainfall required by these crops is 1,500 to 2,000 masl, distributed throughout the year with a dry period of two to three months, in which there is a vegetative rest period, after which flowering takes place. The average optimum temperature for the development of the crop is between 19 to 22°C, with minimum of 16°C and maximum of 25°C, higher or lower temperatures produce diseases in the plants and bad quality in the grain. Soils for coffee cultivation should have a porous space of 60%, of which half should be occupied by air when it is under humid conditions, preferably they should be deep, permeable and of frank texture, so that the roots penetrate with ease, they must be well fertilized and with a pH of 4.5 to 5.5 (INFOAGRO, 2017).

1.5.2. Coffea Canephora

It is native to quite low altitudes and the wetter regions of the West Coast of Africa, although the best *Robusta* coffee is produced in Tanzania at an elevation of 1,200 masl, the crop can thrive from zero masl. It requires an annual rainfall of 900 to 2,500 mm, ideally the precipitation must be distributed evenly for nine months, followed by three months of drought, which induce flowering for the next season, the crop develops in temperatures between 18.3 and 26.7°C with ambients, the soil for coffee should be franc to clay loam, with good drainage, with approximately 1m depth, rich in organic matter, and with a slightly acidic pH of 5.5 to 6.5 (Saucedo-García, et al., 2014).

2. International Market

In the international market, four categories of coffee are distinguished, according to the type of grain, in descending order with respect to quality and price are: soft Colombians, washed Arabica beans, produced mainly in Colombia; other softs, Arabica beans, of which main producers are Mexico and Central America; natural Brazilians, unwashed Arabica beans, from Brazil and other South American countries; and robust, produced in Africa, Asia and some South American countries, in coffee production, South America has leadership due to the soil, climate and altitude conditions, which are very favorable for the crop (EUROMONITOR, 2017).

2.1. International Coffee Production

The main coffee producers in the world (Figure 2 and 3) are Brazil, Vietnam, Colombia, Indonesia, Ethiopia, Honduras, India, Peru, Uganda,

Guatemala, Mexico. In the period 2017/2018, it is estimated that the production was 158.93 million bags of coffee, compared to 157.69 million bags of the previous year of which 62.2% is Arabica coffee and 37.8% robust (ICO, 2017).

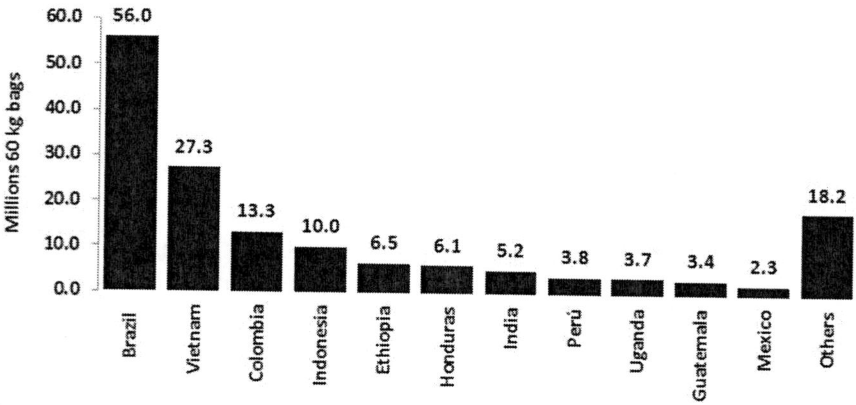

Source: USDA, 2017.

Figure 2. Main coffee producing countries (Millions 60 kg bags, equivalent in green coffee) 2016/17.

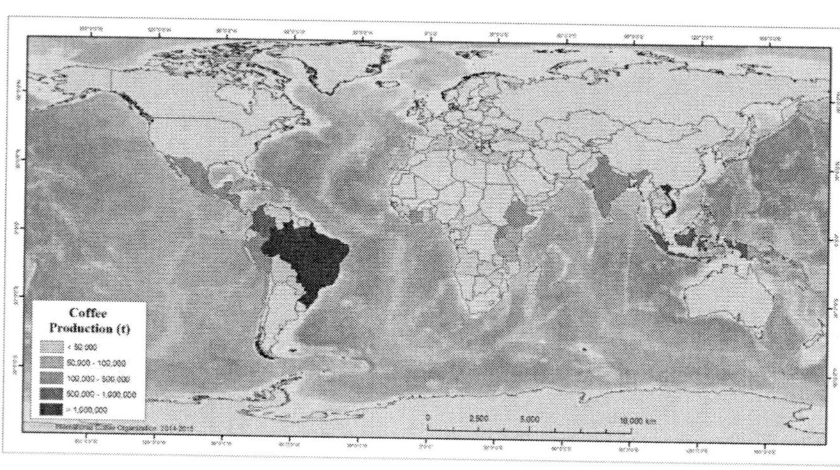

Figure 3. World coffee producer (FAOSTAT, 2017).

The total coffee exports made in February 2015 was 8.6 million bags, 10.2% less than in February 2014. This places the total of world exports made in the first five months of the coffee year (October to February) in

43.4 million bags, 1.2 million bags less than those of that period in 2013/14, of the volume exported 85.9% corresponded to green coffee, 2.5% to roasted and ground coffee and 11.6% to soluble coffee (Central America Data, 2018, ICO, 2017, FIRA, 2016).

2.2 International Consumption

An initial estimate of world coffee consumption in the calendar year of 2014 was 149.3 million bags. This represents an average annual growth rate of 2.3% in the last four years. Much of that growth occurred in emerging markets, with an average of 4.6% since 2011. There has been strong demand, especially from the Russian Federation, the Republic of Korea, Algeria and Turkey, with an increase in demand of the exporting countries on average of 2.6%. Brazil is by far the largest consumer of coffee among the exporting countries, with 20.8 million bags in 2014, followed by Indonesia (4.2 million), Ethiopia (3.7 million) and Mexico (2.4 million) (ICO, 2017).

3. National Coffee Market

Mexico produces coffee varieties classified among the best in the world, the generic variety that is produced in Mexico is the "Arabica," which is classified within the group of "other softs." The Coatepec, Pluma Hidalgo, Jalatenango, Marago and Natural de Atoyac varieties stand out for their quality (Canet-Brenes et al., 2016).

3.1. National Production

Coffee represents a national strategic activity; it employs more than 500 thousand producers, in about 690 thousand hectares of 12 states and 391 municipalities; it involves exports for 897 million dollars per year and is the main producer of organic coffee in the world, allocating 10% of the surface to this activity; In addition, it directly and indirectly links about three million people and generates a market value of around 20 billion pesos per year (ASERCA, 2018, CEDRSSA, 2016, SAGARPA, 2017).

The coffee regions are concentrated in four zones: the slopes of the Gulf of Mexico, which include the states of San Luis Potosí, Hidalgo, Puebla, Mexico and Veracruz. The slope of the Pacific Ocean which covers the states of Colima, Guerrero, Jalisco, Nayarit and part of Oaxaca, the Central-North zone and the Soconusco in Chiapas in the Mexican southeast belong, comprising a large proportion of the state of Chiapas, highlighting the production of organic coffee, which is highly demanded in the North American and European markets, which together includes 398 municipalities in the 12 producing states (SIAP, 2017).

Currently, the cultivation of coffee in our country is developed in 15 states, with an area of cultivation in the 2014/2015 coffee cycle of 727,385 hectares, of this surface 666,375 hectares were harvested, out of which 90.3% was concentrated within five entities: Chiapas (37.2%), Veracruz (20.3%), Oaxaca (17.6%), Puebla (8.9%) and Guerrero (6.3%), allocating more than 80% of production for export, which generates numerous jobs in the field, besides being the main source of income for many small producers and 17 indigenous groups in the country, being approximately 95.0% of the national coffee production of the Arabica species and 5.0% corresponds to the Robusta variety (Valencia-Sandoval and García-Velázquez, 2017; FIRA, 2016) (Figure 4).

Mexico ranks first in the world in the production of organic coffee, being the main producing states: Chiapas, Oaxaca, Veracruz and Puebla, with a production of 350,000 bags of 60 kilos of green coffee, organic coffee; requires strict certification standards, because consumers are concerned about health issues and preservation of the environment (CERTIMEX, 2014).

However, the negative effects of coffee rust have also affected the profitability of organic production, which is why many producers are opting for other certifications of sustainable production with recognition in the world market, such as the Rainforest Alliance, which works through a network by promoting sustainable agriculture to develop best practices for agriculture, incorporating social and environmental aspects, Fair Trade, which requires a certification, which explicitly prohibits the exploitation of children and promotes gender equality, as well as Max Havelaar etiquette which encourages buyers to commit to buying coffee at a minimum price even if world prices are lower than this maximum limit (FIRA, 2015, CEDRSSA, 2016, Medina-Meléndez et al., 2016, 40-41).

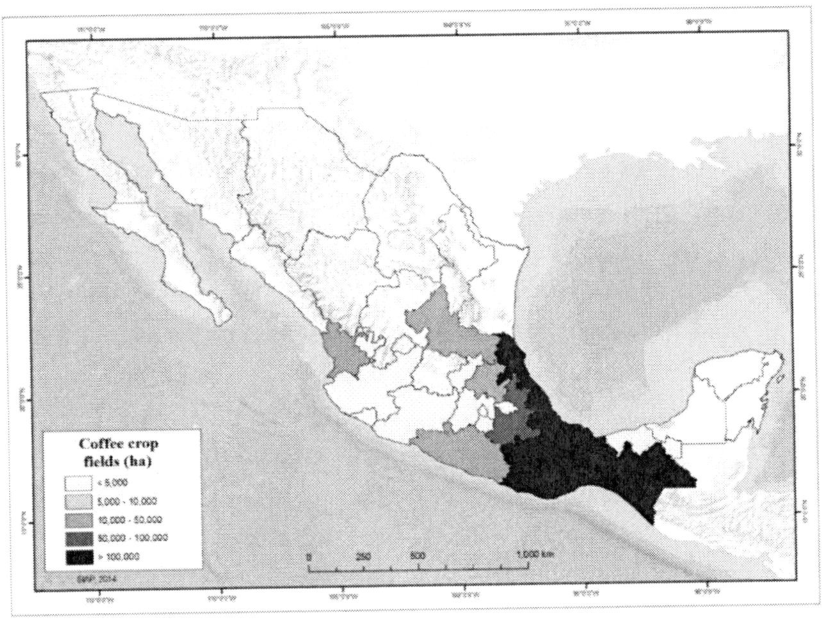

Figure 4. Coffee production in Mexico (Data from SIAP, 2017).

3.2. National Consumption

One of the most consumed beverages in the world is coffee, statistically 2,500 million cups are served in a day (Pardo-Lozano et al., 2007, 228-230). In Mexico, between 0.8 and 1.4 kg per year of coffee per person is consumed, which is approximately 2.6 cups per day (Barreda-Abascal et al., 2012).

Although coffee is a non-caloric, bitter-tasting drink (Aguilera et al., 2009, 178-179), coffee consumption in Mexico continues to increase, in 2016 it grew at an average annual rate of 2%, reaching the 2016 cycle in 2.35 million bags of 60 kg, the import of coffee in Mexico, doubled with respect to the previous year, importing 39,492 tons of coffee, equivalent in green coffee to: 78.8% unroasted coffee, 9.5% roasted coffee and 11.7% soluble coffee and extracts. The above, means an increase of 97.3% with respect to the volume imported in 2013, imports during the first months of 2015 increased at an annual rate of 34.0%, to reach 19,765 tons, equivalent in green coffee (FIRA, 2016; CEDRSSA, 2016).

One of the causes of the increase in coffee consumption is that the myth about the damage of coffee to health has begun to disappear (Fonseca-García et al., 2014, 228-235), thanks to advertising campaigns and favorable opinions in medicine, coffee is now seen as a product that provides status, improving the perception of coffee consumed locally. According to USDA, 64.9% of coffee consumption in Mexico corresponded to soluble coffee and 35.1% to roasted and ground coffee, due to the fact that Mexicans prefer to consume coffee in their homes and mainly use soluble coffee, although the impulse of roasted and ground coffee is driven daily by the new generations especially in the middle and upper classes, which prefer quality and taste rather than price, as well as the expansion of the coffee shops, being Guadalajara and the City of Mexico the cities where the coffee culture has more development (USDA, 2017, INFORURAL, 2018, EUROMONITOR, 2016).

3.3. National Prices

The trend is an increase in the price of coffee, mainly parchment which depends on the variations of the price of this product in the rural environment, which in turn depends on the international price, this volatility of the price of the international markets affects a large number of peasant families in developing countries that see how coffee growing in one year can generate almost all of the family income and the next year produce large losses (Benítez-García, 2015, 193-197).

In Mexico, the price of coffee (expressed in Dollars per Pound) is governed by the "C" contract on the New York Stock Exchange (ICE), in which the price of Arabica coffee produced in Mexico is very high. similar to the trend reported by the indicative price of the International Coffee Organization (ICO), mainly in the "other softs" group, only in October 2016, the price of Mexican coffee stood at $ 3,597 per ton of green coffee, at the New York Stock Exchange, which was an increase of over 16.9% compared to October 2015 (ICO, 2017).

The fact that Mexican coffee is regulated by an international organization is contrasted because coffee cultivation in Mexico is mostly carried out by small producers and family businesses. There are several aspects that cause variability in price, among these, the demand and supply, consumer preferences, weather conditions, the level of inventory, speculation in the prices of the "ball" and the position of the producer countries, among others, which generally imply an unfair exchange against small producers compared to intermediaries and trading companies (CEDRSSA, 2016).

That is why organizations such as FIRA, CEDRSSA, SAGARPA among others look for programs to add value to coffee by differentiating products, satisfying the demand of consumers who are willing to pay a premium for better quality and better product benefits, among recognized programs such as the certification of organic coffee, ecological brands that care about consumer health and the environment, the Max Havelaar etiquette or Fair Trade which seeks fair trade and prohibits child exploitation (CERTIMEX, 2014).

3.4. International Exchange

Total coffee production during 2016 was 824.08 thousand tons, national consumption was 809.84 thousand tons, exports were 79.92 thousand tons, and imports 65.67 thousand tons, being the months of May and June the ones with the largest amount of imports and between March to May where coffee is most imported (SIAP, 2017).

Out of the total national coffee exports that represent 8.98% of total production, the main consumer is the United States with 53.8% of exports, the rest is destined countries of the European Union block mainly to Belgium, Germany and others like Japan, Cuba and Canada (SAGARPA, 2017, SIAP, 2017).

National consumption is completely satisfied with domestic production, however coffee production fell by 43.51% in the last decade, which foresees an increase in imports mainly from Brazil, Honduras and Vietnam, "while exports" (SAGARPA, 2017; SIAP, 2017).

4. Social and Economic Conditions of Producers

Despite the importance of the national coffee sector, it faces a crisis due to the international price of coffee which dropped by 24% between December 2014 and December 2015. Despite the fact of having a slight increase during the quarter from September to November 2016 (Figure 5), low prices have impacted the income received by producers and day field workers during the 2015/2016 harvest, and even generating situations where production costs surpassed income from the sale of these crops, the impact for small producers and workers in some regions remains very strong, causing plantations being totally abandoned, plus the migration of producers and the dramatic drop in the level of human development in the coffee regions (Escamilla-Prado et al., 2015, 359-361; ICO, 2017).

Coffee in Mexico is a crop where 90% of the harvested area of coffee is located in the humid tropics of Mexico, and in areas of high and very high marginalization. According to figures from the National Council for the Evaluation of Social Development Policy (CONEVAL), of the 350,000 coffee production units, 56.6% are indigenous language speakers. This population owns 43.9% of the area with coffee and harvest 40.3% of the volume of production, we will hardly find a greater importance of the indigenous population in the production of any other crop, additionally, the coffee-growing municipalities are characterized by the poverty of its population, 7 of every 10 inhabitants live in conditions of high and very high marginalization. Under this context, there is a concern over impoverishment in coffee regions due to the fall in production, which causes lack of work and therefore, income reduction (Espinosa-García et al., 2016, 2012-2016; SAGARPA, 2017).

The coffee-growing municipalities are characterized by the conditions of marginalization and poverty; 86.3% of the inhabitants of the municipalities dedicated to the cultivation and collection of coffee live in conditions of marginalization; it has been observed that the production of coffee is associated with poverty, which it is somewhat contradictory if we look at the huge amounts of foreign currency generated by coffee for the country (Ortega-Hernández and Ramírez-Valverde, 2013, 183-184).

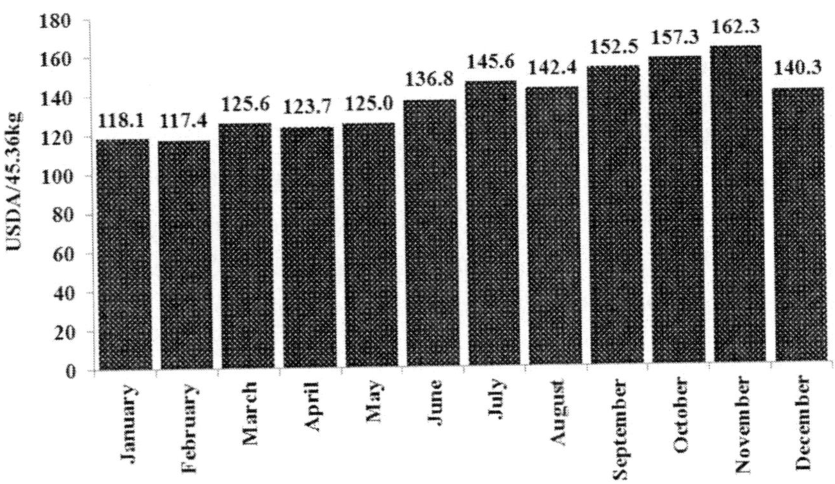

Source: Intercontinental Exchange (ICE).

Figure 5. Price of Contract "C" of Arabica Coffee averaged 1st and 2nd position, Year 2016.

It is estimated that in Veracruz about 50% of the population in the 10 main coffee producing municipalities in the country live in poverty (FIRA,

2016). It is also a crop with important "technological regressions" and without a comprehensive coffee policy; along with the cyclical phenomena of a highly volatile market, the relative technological attraction that is distinguished by the progress in other neighboring countries, which presents a gap of 20 years apart from its primary production and in the transformation and industrialization; this scenario goes to show, the needs for improvements and technological innovations which are required throughout the coffee production chain in Mexico (AMECAFE, 2017), in addition to this, coffee is a plantation that is cultivated in small areas of land by indigenous producers, who live under conditions of marginalization and poverty (Ortega-Hernández and Ramírez-Valverde, 2013).

The processes that shape poverty and marginalization form a precarious structure with few social opportunities for communities, families and citizens, exposing themselves to social vulnerability, for example, housing, people living in unworthy and unhealthy homes, that lack of drinking water, drainage, electric power among other services, generating higher risks and a higher chance of acquiring illnesses throughout their lives, which prevent children from going to school among other important deprivations they have to face (CONAPO, 2015).

Without the appropriate scope on regulations or recommendations which deal with the use of different varieties of coffee, the vast majority of producers tend to mix several varieties in their plots even within the same furrow, and in addition to that, government programs create nurseries where varieties are not regulated, promoting the dispersion of root nematodes between different territories, or other phytosanitary issues. In order to counteract the attacks by nematodes, some producers, especially in the Huatusco region, perform grafts on Robusta coffee roots, which due to its profuse root, withstand the presence of nematodes. This type of strategy of renovation of coffee plantations is oriented to increase production, which in turn, encourages the proliferation of varietal disorder that should be organized by agroclimatic regions, agroindustrial capacities and market supply of various qualities and/or risks of pests and diseases (INIFAP, 2018).

5. Agro Ecosystems of Coffee

Agroecosystems in Mexico overlap in their distribution with important natural ecosystems, reaching 22 million hectares, corresponding to 11% of the territory (SIAP, 2017). Depending on the type of crop, management, extension and many other factors, there has been a different impact of agroecosystems on the biodiversity of natural ecosystems, however the biodiversity remaining in these landscapes also has an effect on agro-ecosystems (Montagnini et al., 2015, 5-20).

National coffee plantations overlap latitudinally with temperate forests, in particular mountain cloud forest (BMM) or cloud forest (Challenger, 1998, 847). This type of forest occupies less than 1% of the total area of Mexico but contains between 10 and 12% of all plant and animal species that exist in Mexico (Sosa-Fernández et al., 2017, 63-66), more than 90% of the area cultivated with coffee is under diversified shade, which contributes considerably to conserving Mexican biodiversity, because as several studies touch on this subject, the vegetal structure of these agroforestry systems is very similar to that of native forests because, the cultivation shade coffee (rustic) can preserve canopy trees and some of the understory that function as shade and therefore can house many species that includes 2,500 species of plants that grow preferably or exclusively in this type of ecosystem, in addition, it houses a high number of endemic species of reptiles (102 species), amphibians (100 species), birds (201 species) and mammals (46 species) being thus a provider of services to society (Escamilla-Prado et al., 2015, 358-359; Ruelas-Monjardín et al., 2014, 36-38).

Despite the importance of shade coffee in Veracruz, dramatic changes in distribution and extension have recently been recorded (Muñoz-Villers and López-Blanco, 2007), which include the reduction of 84% of the cultivated area, the increase in migration, abandonment of plantations, severe environmental impact due to the removal of coffee plantations in exchange of cultivation or urbanization, high incidence of pests and diseases that affect the quality of the grain, low yields and a drastic decrease in the standard of living among rural inhabitants (Ruelas -

Monjardín et al., 2014). All this puts at a high risk the conservation of the agroecosystem of coffee under shade which, as it has been shown, brings about a very important environmental function (Ruelas-Montjardín et al., 2014).

5.1. Coffee Processing Chain

The coffee as a drink is prepared with the seeds after the transformation process (Contreras and Hernández-Martínez, 2008, 9-10) since it has several stages of the transformation chain, which begin once the harvest has been carried out, the purpose of this approach is to separate the grains and have them ready to be roasted, by which the pulp, mucilage and parchment are eliminated. This method is carried out by a wet or dry route process (SAGARPA, 2017, Giraldo-Quintero et al., 2017, 123-133).

5.2. Wet Process

Under the wet process, the pulp is removed to the grain by means of the pulping machine, this being the first physical transformation of the coffee, resulting in a coffee covered by a film of mucilage that still wraps it and that is not soluble in water, so it must be fermented submerged in water tanks for a period of time between 18 to 30 hours, the second transformation process to which all coffees go through is washing, during this phase, mucilage (endocarp) is eliminated, once completed, the grains are collected and placed in large wicker baskets, where they are scattered on huge outdoor terraces, where they are turned over constantly over a period of time until the sun and air dry them accordingly. This process is carried out using the three following steps: pre-drying (wet parchment coffee), sun drying (wet parchment coffee) and machine (dry parchment coffee) until obtaining between 10 and 12% humidity, obtaining what is called, golden coffee (CENICAFE, 2018).

5.3. Dry Process

This method consists of separating the coffee beans from their husks in one complete batch, in order to carry out this operation the whole cherry is dried out, first the cherries that have been collected are sorted out and cleaned, the cherries that are not ripe enough are separated from those that are too ripe and damaged, and then the dirt is removed, soil, branches and leaves, and after that, the coffee cherries are spread out under the overcasting sun, either in large gardens of concrete or brick, this approach can also take place on mats raised to the height of the waist on trestles, as the cherries are left to dry, they are raked or turned by hand to dry homogeneously, it usually takes about four weeks for the coffee cherries to reach their maximum level of moisture content 12.5%, depending on atmospheric conditions, on larger coffee farms there is sometimes a machine that performs the drying in order to accelerate the process after the coffee has been dried under the sun for a few days, the drying process is the main stage, because it affects the final quality of the green coffee, when the drying process is completed, the peel becomes brittle and of a dark brown appearance, in addition to shaking approach which even lets you hear the peculiar sound caused by loose seeds, right then, it's time to screen and to peel, to move by threshing, which is where the green coffee beans are selected (Spanish Coffee Federation, 2018).

5.4. Coffee Yield

The proportion of the components of the coffee fruit and its physical characteristics can undergo modifications both in the stage of the crop and in the processing, which will be reflected in the coffee yield ratios (CENICAFE, 2018). 60 kg of green coffee represent: 120 kg of dried cherries, 75 kg of parchment and 50.4 kg of roasted coffee (ICO, 2015).

6. Coffee Roasting

A good cup of coffee should gather basic qualities such as aroma, body, acidity, bitterness, and finesse, these properties are developed according to the characteristics and management of the crop and to a large extent to the treatment which the grains received during the overall process (CENICAFE, 2018).

The green coffee bean has no smell or taste and is only of a pale green shade appearance which will change to a dark brown grain once consecutive process is applied. All the aroma and flavor that we enjoy in coffee is created by toasting of the grain (ICO, 2017).

The degree of roasting is indicative of the intensity by which the grain was roasted and varies according to the specific characteristics that one wishes to obtain on the final stage of the product to satisfy the different tastes and preferences of the consumer, in Mexico three types of toast are commercialized, which are measured mainly with colorimeters (PROY-NMX-F-173-1999) the roasting of the coffee beans takes place by roasting equipment that is heated to a temperature between 210 and 230 ° C, in which the grains are deposited inside rotating drums which in turn initiate the desiccation of the grain where a chemical reaction of the compounds found in the almond takes place, the starch is converted into sugar, the sucrose is caramelized, the amino acids are degraded, the oxidations of the lipids and of the phenolic acids, the reactions and formations of the acids, the proteins are broken down and the whole cellular structure of the grain is altered, the heating process releases the coffee oil, or what is called "coffee," which is the essence of coffee, the color and production of volatile compounds and melanoidins are generated and, in addition, hundreds of other substances that form are detected in the various reactions, through heat, during roasting (Marín-García and Puerta-Quintero, 2008, 10-25; PROFECO, 2001).

There are coffees with a wide range of tan color that cover everything from the lightest, such as New England coffee, to the darker ones, such as Italian or espresso coffee (Gotteland and de Pablo, 2007, 110-112).

7. MARKETING OF COFFEE IN MEXICO

The commercialization of coffee in Mexico is an extremely complex process, since local prices refer to the international market rates (expressed in dollars per pound) as per the Contract "C" at the New York Stock Exchange (ICE) and that itself, in the context of a liberalized market, intermediation processes has been the leading role, which makes the distribution of values along the coffee production chain unequal, together with the fact that coffee consumption is scarce in most of the producing countries that grant greater added value in the importing countries (ANACAFE, 2014). The Mexican producer has a serious problem due to the lack of information from the markets abroad and the conditions thereof, he goes about ignoring the levels of fluctuations in prices related to the international market and if he has access to some form of information, he doesn't know how to match prices at its equivalent price within his local market or territory (Coutiño-Puchuli et al., 2017, 45-49).

Another factor that adds complexity to the commercialization process in Mexico, which only makes about 7.0% of primary marketing (produced sales), is in the form of green or golden coffee. This does not include precise information about sales of roasted and ground or soluble coffee that has been directly linked by its producers, but estimates that people do not exceed the 3.0% mark of the total volume traded. According to data from the SICN, there are 230 active marketers in Mexico who locate the beneficiaries, exporters and formal marketers of the grain (AMECAFE, 2014, SICN, 2017).

Of the national production of coffee, 60% is exported, which is why marketing requires the different channels in which various agents intervene, the main modalities being the most common forms are shown below (Figure 6).

The three main channels of sale of coffee are retail making up for 70% of the total sales, where the type of coffee with the highest sales is soluble, in total coffee distribution: Independent stores represent 32.0%, Discount wineries 23.0%, Hypermarkets 19.0%, Price clubs 10.0%, Supermarkets

9.0%, Coexistence stores 6.0% and Others 1.0%, Foodservice, whose importance and growth is due to the growth of emerging new coffee shops and the change in the concept of coffee in Mexico, ground coffee is the type of coffee preferred by the foodservice followed by soluble coffee in areas that do not have a coffee machine available, coffee distribution in this sector is contributed to: restaurants 55.0%, coffee shops 23.0%, fast food chains 14.0% and others 8.0% and finally the institutional channel, which is sometimes confused with the foodservice, the institutional channel has clearly involved in large companies such as hotels, public sector offices and even businesses that require it with large staff numbers. The distribution of this channel represents in total by: public sector offices 24.0%, private sector offices 23.0%, hotels 19.0%, educational centers 18.0%, hospitals 7.0% and other 9.0% (Euromonitor, 2015).

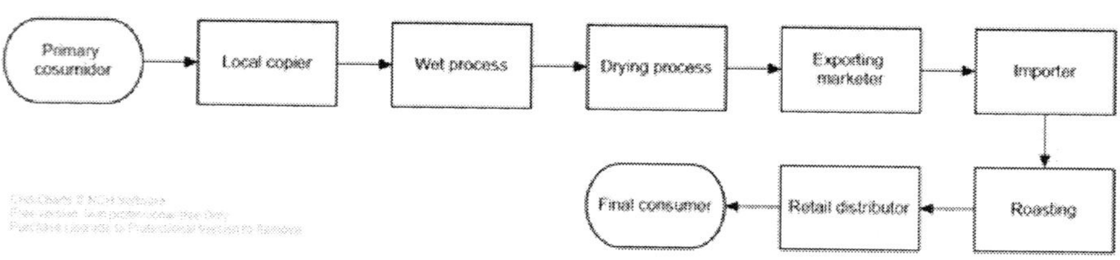

a) International market

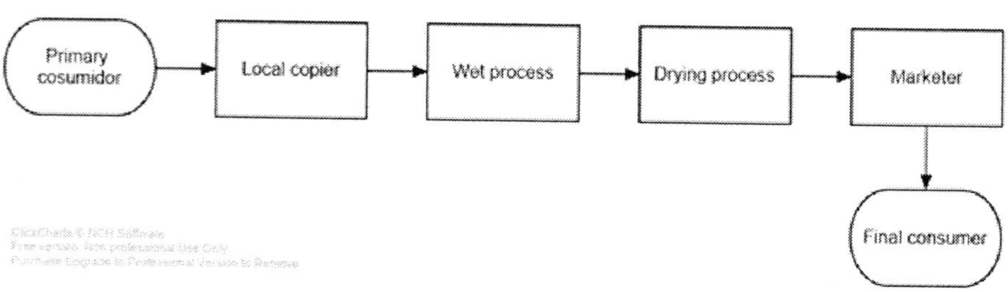

b) National market

Figure 6. Main coffee marketing channels (Data collected by Figueroa-Hernández et al., 2016, 68-69).

7.1. Socioeconomic Conditions of Coffee Companies in Mexico

There are two decaffeination companies (CAFIVER and DESCAMEX), 10 solubilization companies, the only freeze-dried coffee factory in Mexico and Central America (CAFESCA), 16 large and medium sized roasters and an unknown number of small roasters that need to consolidate and professionalize their activities. It is estimated that about 60.0% of the national market for soluble coffees is served by Nestlé; while other companies with a national presence such as GRUPO MARINO, CAFINCO, SABORMEX, INTERCAFÉ, among others, strongly compete for the consumer preference with recognized brands. There are other companies such as GESOL, CAFIVER and KATOEX that have specialized in the maquila of soluble coffees for other private brands. This link in the productive chain is one of the most consolidated through organizations such as The National Association of the Coffee Industry, A.C. (ANACAFE) that brings together beneficiary, industrializing and exporting companies, as well as companies dedicated to the operation of cafeterias; the Mexican Association of Coffee Exporters, A.C. (AMEC) groups the main exporters (natural and legal persons) and Section XX of Roasters and Coffee Mills of the CANACINTRA (AMECAFE, 2016).

8. Coffee Residues

During the industrialization of coffee, most of the fruit is not used or marketed, which generates a considerable amount of waste, which pollutes the air, soil and water, an example are the discharges of wastewater, which constitute foci of vectors and the accumulation of waste, which alter the quality of air, that is why the by-products of coffee have been studied to find industrial, energy and food applications, providing an added value when used in the production of ethanol, compost, fertilizer, paper, unicellular protein, as well as in the cultivation of edible fungi and in the extraction of active metabolites, so that contamination is avoided this way (Peña-Martínez et al., 2016, S1175-S1177; Noguera and Posada, 2015, 1-7; Rodríguez-Valencia, 2013, 79-80).

8.1. The Pulp

The pulp is formed by the exocarp (epidermis) and part of the mesocarp (mucilage); the color of the epidermis varies from green to red, depending on the variety of coffee and the degree of ripeness of the fruit, with the pulp being the first product obtained in the processing and representation stage, on a humid basis, around 43.58% of the weight of fresh fruit (Montilla-Pérez et al., 2008, 126-139).

The composition of the coffee pulp, is composed of a dry matter content of 87.51%, humidity 76 to 80%, crude fiber 20.23%, crude protein 11.20%, ash from 11 to 12%, cellulose 29.50%, lignin 16.90%, hemicellulose 4.60%, tannins 3.70%, caffeine 0.75%, nitrogen 1.30%, reducing sugars 17.30%, Brix 3.59° and total sugars 18.49% (Rodríguez-Valencia, 2013a).

Out of the fresh fruit from the coffee, the pulp represents 42% of the total weight, which is why it is the most voluminous by-product of the wet processing, of every two tons of coffee, approximately one ton of pulp is produced (Duangjai et al., 2016, 10-14).

8.2. Mucilage

When removing the pulp, the grain is covered by the mesocarp, better known as mucilage, which is a layer of translucent and viscous consistency tissue that represents 15.5 to 22% of the weight of the ripe fresh fruit, the mucilage measures approximately 0.4 millimeters thick, it is rich in reducing and non-reducing sugars as well as pectins, generated from the calcium pectate that during the maturation stage of the coffee bean, the mucilage is strongly adhered to the endosperm of the seed, which due to the colloidal nature of the mucilage is indispensable to be removed to facilitate the drying of the grain, avoiding unwanted fermentations and preserving the quality of the coffee (Blandón-Navarro et al., 2013,10-13).

The mucilage and the fermented coffee honey depend on the coffee production, so they are generated in a discontinuous way and one gives the variables of the quantity of the mucilage at the ripeness of the fruit, in mature and fresh fruits the mucilage content is approximately of about 10.4% in relation to the weight of the fruit, therefore for each ton of cherry coffee that is processed, between 80 and 140 kilograms of mucilage can be obtained, depending on the maturity and the amount of water used in the process of the demucilage (Puerta-Quintero and Frías-Arias, 2011, 24-27), this represents 28% of contamination as a consequence of the generation of the wet process (Rodríguez-Valencia, 2013, 80-82).

8.3. Husk

The brown husk also called cisco corresponds to the endocarp of the coffee fruit (parchment) and which covers the seed, it's a cartilaginous yellowish-white layer with a certain degree of opacity of approximately 100 μm in thickness that represents 4.2% of the weight of fresh fruit per kg of fresh coffee fruit, from which 50g of husk are obtained (Rodríguez-Valencia et al., 2013b, 128).

Coffee husk is composed by 5 to 8% moisture, dry matter 92.8%, ethereal extract 0.6%, nitrogen 0.39%, ash 0.6%, bulk density 0.33 g/cm^3 and cellulose 57% (Manals-Cutiño et al., 2018, 172-179).

8.4. Sediment

During the preparation of the drink or during the manufacturing process of soluble coffee, the roasted and ground beans are put in contact with hot water, releasing the phenolic, organic and aromatic compounds, giving the drink the properties which characterize it, it is in this phase where the solid residue known as sediment or spent coffee is produced, which, having no commercial value, is discarded and sometimes used as a compost (Zuorro and Lavecchia, 2012, 49-50).

According to reports from the world coffee organization, the world production of coffee beans exceeded 8 million tons, which indicates that large amounts of coffee waste are generated, even though these residues are integrated of compounds that can have a commercial interest, this plant material does not have a specific defined use and a significant percentage goes to landfills (Castellar-Ortega et al., 2017).

The coffee sediment contains large amounts of detergent fiber, such as cellulose and hemicellulose, approximately 45.3% in dry weight, 21.2% mannose, 13.8% galactose, 8.6% glucose and 1.7% arabinose (Mussatto, et al., 2012, 176-178), 13.6% of proteins, total nitrogen from 8.5 to 13.6% (Belitz et al., 2004, 939-969). The content of caffeine in the coffee sediment varies according to the species, Arabica coffee has approximately 0.2% and Robusta 0.5% (Ramalakshmi et al., 2009, 82-84).

The amount of coffee waste generated represents a contamination hazard, so potential alternatives have been sought for its use in the industry due to its high calorific value of 5,000 kcal/kg, it is possible to use it as fuel in industrial boilers (Hernández-Cázares et al., 2016, 13-14).

9. Potential Uses of Coffee Waste

9.1. Biogas

Biogas, is the gas generated from organic digestion by a population of mixed organisms, under anaerobic conditions, generally this anaerobic conversion of organic matter is produced in three stages that are: 1) fermentative, 2) acid forming and 3) methanogenic (Anunputtikul and Rodtong, 2004).

Among the uses of biogas is the generation of electrical energy, fuel in stoves, substitute for gasoline and diesel (Taramona-Ruiz et al., 2017, 51-53), while waste sludge from biogas production can be used as organic fertilizers (Office of Energy Efficiency and Renewable Energy, 2003).

The coffee pulp is a potential source for the production of biogas, an adequate fermentation causes the pH to be reduced and the formation of CO_2 begins, reducing during the process tannins and polyphenolic compounds, this makes possible the production of enriched methane, because it will only have half the level of CO_2 that biogas produced by other means generally contains (Kim et al., 2016, 5-6).

9.2. Silage

Coffee pulp has been used mainly as silage in animal feeding systems, a subject on which various investigations have been carried out on the process, the chemical composition and the cost of the silage, reaching the conclusion that in order to use the pulp of coffee for animal feeding, it is necessary to improve its nutritional value without increasing the cost of the final product, being a process with great prospects of being applied successfully (Noriega-Salazar et al., 2008, 413-417). According to Rathinavelu and Graziosi (2005) the inclusion of silage coffee pulp in the diet of some animals helps reduce the costs of milk and meat production, especially in developing countries.

9.3. Edible Fungi

The cultivation of edible fungi is an important alternative to satisfy the nutritional needs of the population, agricultural development and formation of agroindustries in many parts of the world, in addition to using agroindustrial waste for their production (Romero-Arenas et al., 2015, 1237).

Among the agro-industrial wastes that have potential among the production of edible fungi are the by-products of coffee, however, due to the content of tannins and caffeine, their use in animal feed and in bioprocesses is limited, so detoxification plays an important role in its use, studies have shown that in detoxified coffee pulp, 73% is used in fungal crops, and the rest can be used as a compost to produce a fertilizer for soils (Bermúdez-Savón et al., 2018, 249-253).

9.4. Biofuels

Biofuels are the fuels of biological origin obtained from organic waste, which can be classified according to their type and consistency: 1) solids such as wood, tablets or pellets, 2) liquids such as bioethanol and biodiesel and 3) gaseous like biogas (Lede, 2008).

The problem of climate change and the search to promote sustainable development, has made biofuels increased their international importance, among these the production of biodiesel is reaching the most relevance because it is a good alternative for fossil fuel (Montero-Alpírez et al., 2017, 78-81).

An alternative for this is the use of waste from the coffee agroindustry, an example of this is the biodiesel produced from the coffee waste, Kim et al., (2016) used coffee sediment to produce biofuels such as biodiesel, ethanol and biogas using physicochemical or biological processes with good results.

The mucilage of coffee, because of its high content of sugars, is an excellent option for the production of bioethanol and other alcohols, Blandón-Navarro et al., (2013) mentioned that the mucilage of coffee added with 1% yeast and fermenting at 35°C produces 98.6% ethanol yields of 57.90 mL of mucilage and a residue of the distillation of 44.42% of the initial volume, which represents an excellent opportunity for economic development for producers.

Another by-product of coffee with potential for the production of biofuels is husk, which has no value as a fertilizer, so it is partially dried in the sun and used as a solid fuel (Dicovskiy-Rioboo et al., 2014, 11-12).

Sustainable development includes the use of renewable energies, thanks to the fact that it has a low environmental impact, reducing dependence on the use of fossil fuels (Montero-Alpírez et al., 2017), that is why the coffee husk mixed with another type of biomass, is already used as auxiliary fuel in gas boilers of some processes (Hurtado-Rengifo, 2016, 29-30).

9.5. Compost

One of the technologies that has been implemented to minimize the contamination caused by coffee byproducts, has been to take advantage of the nutrients they have to make compost and be applied as organic fertilizer, studies have revealed the coffee pulp combined with chicken manure produces excellent compost quality and greater microbiological diversity (Hernández-Cazares, 2016).

That is why composting and vermicomposting provide a solution for the agroindustrial waste of coffee, creating fertilizers rich in nutrients that increase the productivity of plants. Another use for coffee pulp is the production of vermicompost, for every 10 tons of fresh pulp, one ton of vermicompost can be generated (National Federation of Coffee Growers of Colombia, 2018).

9.6. Alcoholic Beverages

Alcoholic beverages are made from the fermentation of organic materials rich in sugars, alcoholic fermentation is a complex chain where intermediate substances are produced, produced by a complicated enzymatic process, which acts on sugars forming alcohol, anhydride, carbon dioxide, glycerin, aldehyde, acetic acid, succinic acid, butylene glycol and acetoin (Arguedes-Gamboa, 2013).

The coffee pulp and sediment are easily fermented with yeast or lactic acid bacteria that produce alcoholic beverages and vinegars, in addition, these materials contain antioxidants (Rodríguez Valencia, 2013, 81-83), this is important because the phenolic compounds help to define the sensory properties and other attributes in good quality alcoholic beverages such as wines (Tian et al., 2009, 833-834).

These properties of the coffee byproducts, together with the organic acids present, make them a quality raw materials to produce a carbonated beverage, minimizing environmental contamination since they offer the consumer a refreshing, pleasant product with antioxidant benefits (Peña-Martínez et al., 2016, S1175-S1177).

10. Prospective of Coffee as Agro-Industry

Currently, national coffee requirements are met, despite the fact that production has fallen by 43.51% in the last decade, which is why imports from countries such as Brazil, Honduras and Vietnam have increased. In the 2016 of the production in Mexico of green coffee, 51.95% was destined to the national market (SIAP, 2017).

The national consumption of coffee is increasing, so it is necessary to look for trends that require agribusiness to improve their processes, along the value chain, from production and processing, to the presentation and marketing of each and every one of the agri-food products. These tendencies are engaging visions of the future built on what is happening nowadays. They are not mere occurrences or good wishes: they come both from scientific and technological changes, and from the evolution of the dynamics of a society (Valencia-Sandoval, 2015, 242-245).

These trends should consolidate the import and export of Mexican coffee as well as improve production and economic life of producers, considering that one of the important points in the trends of the global agri-food industries is to implement new technologies in production, processing and care of the products and their derivatives, the great challenge of the agri-food sector in the next 15 years is to give satisfactory answers to an increasingly demanding consumer with greater knowledge, which demands more information, quality and safety in the food that will be consumed, this situation applies directly to the national coffee agri-food industry (ILO Foundation, 2017).

Among the strategies for the development of coffee agribusiness are aimed to promote integrated phytosanitary campaigns in the Central

American region, by intensifying the coffee replanting programs, promoting innovation and transferring of sustainable technology, promoting productive diversification, boosting the systems of sustainable production with good practices that maximize the quality of the product, implementing marketing promotion campaigns, based on the development of micro-regions, providing sufficient financial backing needed for the collection, marketing and renewal of coffee plantations, as well as for product capitalization, thus consolidating and promoting systems for the use and reuse of coffee byproducts, promoting at the same time, organic production systems, sustainable technology certification processes, distinctive quality as well as specialty, campaigns to promote Mexican coffee targeted for the development of and growth of Mexican companies (SAGARPA, 2017; EUROMONITOR, 2017; FIRA, 2016).

10.1. Designations of Origin of Coffee

A consequence of international commercial activity, the concern for the inclusion of trade issues in intellectual property in trade agreements, among these issues reference is made to distinctive signs, trademarks and geographical indications and specifically their protection (Pérez-Akaki and Pérez-Tapia, 2011, 45-49).

For this type of products, the place of origin is the most important, since they are endowed with unique characteristics that distinguish them from the rest, as well as having a demand in the market that guarantees their continuity over time. Geographical indications are necessary among today's market economy, because they represent a way to lessen the asymmetries of information presented, since they label a product and create a reputation which greatly allows to differentiate itself from one another (Vidal, 2017).

The DO must be constituted by the name of a place or geographic region of the country, which designates a product originating from that geographical region, whose product has characteristics and special characteristics for products of the same class or species that are exclusively due to the geographical environment, included in the latest natural factors such as the characteristics and composition of the soil, temperature, humidity, height above sea level and human factors including tradition and custom, specialization in a specific art and the use of special processes (IMPI, 2018).

To achieve the recognition of protection of a DO, a prior declaration to the authorities of the country where the protection is recovered is not sufficient, the link between the origin and the quality of the product, and the effort to obtain the DO, is also demonstrated. This is a form of legal protection that guarantees that customers and consumers are identified with this law, starting, of course, with the origin of a specific origin. In this way, the consumer can trust whatever he is buying, and the producer has been rewarded by his efforts to produce a product of superior quality that meets the standards of the denomination (CENICAFE, 2018).

Analysis performed on coffee production in the world has observed that in Mexico some regions stand out in Veracruz such as Coatepec, Córdoba, Huatusco and Orizaba, the Soconusco in Chiapas and Pluma Hidalgo in Oaxaca, although there is also specific mention about the high quality of the grain produced in Colima and Uruapan, Michoacán (Pérez-Akaki and Pérez-Tapia, 2012).

In Mexico, the protection of a DO is initiated by means of the declaration issued for this purpose by the Mexican Institute of Industrial Property, containing in this statement the elements and conditions that constitute and motivate it. Likewise, the Mexican State will be the owner of the protected designation of origin in accordance to the provisions established by the Industrial Property Law and corresponds to the Mexican Institute of Industrial Property, in accordance with the powers conferred by the Law on the subject to authorize the use of protected appellations of origin, on the other hand, the use of a DO without the corresponding authorization is considered illicit, including cases in which it is accompanied by indications such as gender, type, manner, imitation or others that tend to confuse consumers or imply unfair competition (IMPI, 2018).

10.1.1. Chiapas

The Chiapas Coffee protects the green or roasted-ground coffee, of the *Coffea arabica* species, cultivated and produced in the State of Chiapas and belonging to the family of the *Rubiaceas*, by means of the denomination that is given to the coffee that is produced in different zones of the State, in attention to the optimal geographical conditions and culture that it possess, with rich volcanic soils, suitable climates that characterize it for having humidity throughout the year, and a great diversity of excellent temperatures, which give it characteristics as high acidity, intense aroma, pleasant taste and appreciable body, thus reaching an excellent quality that meets strict conditions of excellence, taking into account the tradition of cultivation that has characterized the State, the production areas of Chiapas are constituted by the regions of Ocozocoautla, San Cristóbal de las Casas, Copainalá, Comitán, Ángel Albino Corzo, Pichucalco, Bochil, Palenque, Ocosingo, Yajalón, Motozintla and Tapachula (IMPI, 2018).

The Mexican Institute of Industrial Property, based on articles 6o. Section III, 158, 159, 163, 164, 165, 166, 167 and 168 of the Industrial Property Law, published by the Official Journal of the Federation on August 2, 1994, proceeds to the publication of the General Declaration of Protection of the Denomination of Origin "CAFE DE CHIAPAS," presented before this Institute (Extracted from: General Declaration of Protection of the Denomination of Origin Café Chiapas, 2003).

"CAFE CHIAPAS" contemplates the qualities of raw washing and height, which cover the bulk of the exports, but we can also find in a smaller quantity those of extra-raw washing and strictly height, within the main varieties are those of high size as Typica, Maragogype, Bourbon and Mundo Novo, which can have an average population per hectare of 1,500 to 2,000 plants; and low varieties such as Caturra, Garnica, Catuai, Catimor and Pacamara, which can have a population of 2, 500 to 3, 300 plants per ha (General Declaration of Protection of the Denomination of Origin Café Chiapas, 2003).

10.1.2. Veracruz

Café Veracruz is the name given to coffee that is produced in various areas of the State of Veracruz. This coffee arises from the particular combination of its deep volcanic soils; the climate characterized by its high humidity throughout the year; and its cloudy winters, which, together, specify the high acidity, the intense aroma, the taste of spices and the appreciable body that characterize this Mexican coffee, the altitude of its plantations and the deep-rooted tradition of fermenting fresh coffee pulp, guarantee in the region coffee an unparalleled acidity that, added to the physical characteristics of the grain and the qualities of the drink, give Veracruz Coffee an optimum quality nationally and internationally recognized (IMPI, 2018).

"CAFE VERACRUZ" is classified in strictly height, height, extra prima, raw washing or snail, within the main varieties it comes from coffee plants of the *Coffea arabica* species among which the predominating varieties are: Typica, whose cultivation is carried out by 22% of the total producers of the State, Bourbon, Caturra and Garnica in a 19% each, Mundo Novo and others that constitute the remaining 21%, are grown under shade in regions with optimal ecological conditions for the cultivation of coffee; this is at average temperatures ranging from 18 to 22° C, with minimal risk of frost in winter, and with maximum temperatures during spring-summer below 30°C, well distributed rainfall throughout the year, reaching between 1 400 and 2 000 mm, with a small drought in March and April, which promotes flowering; soils more than one meter deep, with a texture ranging from frank to clay loam, pH from 4.5 to 6.0, organic matter contents of around 7% and altitudes between 750 and 1 250 meters above sea level, at latitude N between 81° 15 and 20° 45 (General Declaration of Protection of the Coffee Veracruz Denomination of Origin, 2000).

10.1.3. Oaxaca

In May of 2003, the Declaration of Protection of the DO Café Pluma was requested from the IMPI, which includes the parchment or green coffee produced along the Pluma region, which is characterized by a pleasant bouquet, exquisite flavor, accentuated acidity, excellent aroma and mild body. It is classified into four blocks: parchment Coffee, Conventional Green Coffee, Specialty Green Coffee and Roasted, for its attributes and specifications of the *Coffea arabica* species from the fruits of coffee plantations mainly of the Typica variety. Cultivated from 700 to 1,500 meters above sea level. The Pluma region comprises 12 municipalities in the state of Oaxaca, which comprise 27% of the Entity's coffee area, but only 20% of the total area meets the height defined in the application before the IMPI, so which still awaits the resolution for the DO of Café Pluma (SAGARPA, 2017).

10.2. Use of Technology and Adequate Financing

The two species of commercial interest of Arabica and Robusta coffee define the production and supply of world coffee, with arabica coffee having the majority market, with Brazil as the main producer with 40% of world production, in this market, Mexico ranks as the sixth producer of Arabica coffee contributing 5% of total international production in the 2014/2015 cycle (ICAFE, 2014).

Coffee production in Mexico is divided into two types: on communal, ejidal and private lands, in states such as Veracruz and Puebla, private property dominates, while in Oaxaca and Chiapas communal property dominates (Amaro-Rosales and Gortari-Rabiela, 2016, 95-97).

That is why it is evident that a strategy for the conservation and improvement of coffee cultivation is based on coffee plantation renewal programs with rust-resistant coffee trees and all the varieties used and permitted in the production of coffees with denomination of origin, under which two problems are taken into account, in one hand the problem of fungal diseases, which have proved to be the main threat caused by climate change, and the second, young plants increase production as well as the cultivation of coffees with denomination of origin increases the price of coffee beans (National Agricultural Planning of Mexican Coffee, 2018).

The establishment of a research center dedicated to coffee in Mexico, with three fundamental objectives, the first to generate resistant varieties to each coffee zone in the country, study and generate solutions to the problem of diseases and pests in coffee trees, the third one, is to offer solutions to the main technical and technological problems as well as training of coffee producers, this has been successfully tested in other coffee producing countries such as Colombia with the National Center for Coffee Research, creating programs that encourage the use of waste of coffee and provide training to producers, this way pollution is drastically reduced, production costs are lowered and it generates added value to the coffee agroindustry (SAGARPA, 2017).

10.3. Added Value of Coffee by-Products

Coffee consumption of specialty or high quality has grown rapidly in recent years, one of its key attributes is its origin, i.e., the coffee region where they were grown and processed, are usually placed in the same markets: fair trade, sustainable, organic, and gourmet also known as premium coffee, for which consumers are willing to pay even higher prices, however, for coffee to be truly special, it must also represent benefits for the producer (Montoya-Ferrer and Montoya-Vázquez, 2011, 97-104). There is a shortage of specialty coffee in the world market, this means that special coffees can be sold at higher prices and allow them to stay out of the current coffee crisis. (Schüßler, 2009, 171-172).

The production of organic coffee and international certifications of sustainable production increases the price of coffee, in Mexico the main labels that are linked to certificates with standards for sustainable production are the following: Organic, Fair Trade, the Sustainable Agriculture Network (known in the market as Rainforest Alliance™) and Bird Friendly (Sepúlveda et al., 2016, 1007-1008).

To belong to these organizations, it is necessary to have self-sustaining farms and conserve environmental services such as the regulation of water flows and their speed and consequent control of runoff, in addition to conserving biodiversity and allowing them to continue to help production. An alternative for this is giving an added value to coffee byproducts, using waste for the production of biofuels, fungi, compost, papermaking, among other uses, lowering the amount of waste and giving added value to by-products (Mussatto et al., 2011, 668-670).

11. BIOREFINERIES

According to the International Coffee Organization, in 2015, 9.1 billion kg of coffee was consumed worldwide, considering that of the total coffee only used in 5% as a beverage, this generates a large volume of biological waste that contributes to environmental pollution (Kovalcik et al., 2018), so there is an urgent environmental need to reduce the environmental impact caused by these wastes, and at the same time, these wastes have huge economic potential in those millions of tons of coffee waste full of compounds of economic and industrial interest, which is available at very low cost or free of charge (Koby, 2015, 97-98).

The production of biofuels is increasingly considering the use of vegetable waste, among which are the residues of the coffee agro-industry, an example of this is Germany, the world's leading producer of biodiesel, setting the pace with the first refinery that obtains wood fuel and the United Kingdom with a sustainable Bioenergy Center, which handles raw material of agricultural waste, with this, the European Union plans to cover 10% of transport energy with biofuels by 2020 (Quintero-Castaño et al., 2014, 261-264).

Recent studies have focused on investigating the composition of the coffee gum and the uses they may have, among the chemical composition it has been found that the coffee gum contains lipids, minerals, proteins, phenolic compounds, lignin, aliphatic acids, melanoidins, polysaccharides, oligosaccharides, amino acids, alkaloids such as caffeine, and volatile compounds, due to this composition, functional properties have been found in coffee grounds such as water retention capacity, oil retention capacity, emulsification activity, emulsion stability, antioxidant potential, adsorbents, additives for polymer compounds, as sanitary landfills and for its nutritional value as a supplement in animal feed (Kovalcik et al., 2018, 3497-3500).

We have also carried out analyses of integral valorization of byproducts as a raw material to obtain antioxidants, biofuels, bioenergy, bioproducts and biofertilizers, creating biorefineries that generate the aforementioned products, by taking into account the biorefineries design, the production scale and a technical and economic evaluation (Janissen and Huynh, 2018, 110-115).

According to a review carried out by Koby (2015), it is estimated that there are more than eight million dollars a year as a potential gain of the waste generated by Starbucks in the United States, when they are used to produce biodiesel and compressed fuels (pellets).

The biorefinery is an industrial facility that seeks the integral use of biomass in a sustainable way for the production of biofuels, energy, materials and chemical products with added value (Muñoz-Muñoz et al., 2014, 13-16).

After the evaluation of the coffee sediment to obtain different biofuels, the biorefinery can obtain different products (Figure 7), for the formation of biogas, they are based on the conversion of hemicellulose and cellulose into fermentable sugars, using a separation selective by means of an anaerobic hydrolysis of carbohydrates, which releases gases such as methane. (Kovalcik et al., 2018, 3497-3500).

The extraction of oils from the coffee gum can be achieved with different methods, among those like the conventional method which uses organic solvents, the Soxhlet extraction, where an extractor and a rotary evaporator are used and the supercritical extraction method where CO_2 is used, after the extraction of oils by means of biocatalysts, an esterification is carried out for the transformation of biodiesel or a hydroxygenation for renewable biodiesel (Haile, 2014, 67-68).

The bioethanol is first obtained by hydrolysis in order to obtain the sugars contained in the coffee residues, then the hydrolysis is performed until obtaining an average of 58.4 g L-1, which are fermented using Saccharomyces cerevisiae as biocatalyst, until reaching an alcoholic degree, after fermentation, approximately of about 10 to 15%, concentrating by distillation to absolute alcohol with a minimum of 9.4% purity (Karmee, 2017, 7-9).

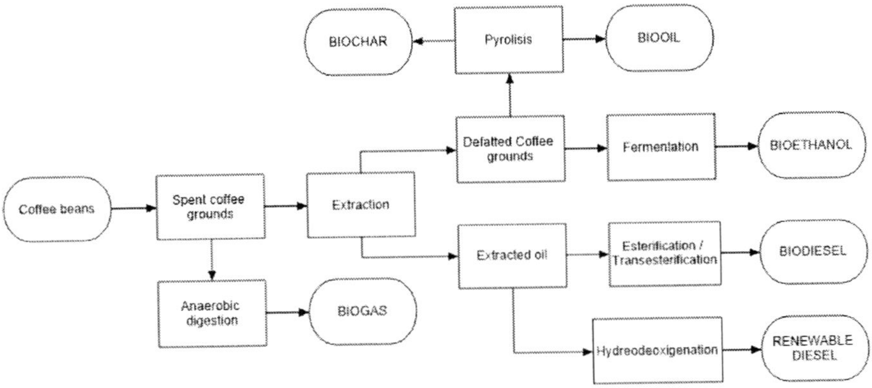

Figure 7. Production of bioethanol, biodiesel, biooil and biogas from spent coffee grounds (Data from Karmee, 2017).

According to Vardon et al., (2013) Bio-oil and fuel can be extracted from the residues of the hydrolysis of coffee grounds by means of a pyrolysis.

The purification and fractionation of coffee residues can be a more efficient and meaningful way to provide an economic value to these wastes, accepting waste materials as a valuable source of organic and inorganic substances that can be incorporated into new projects. Scientific investigations (Kovalcik et al., 2018, 3497-3500).

CONCLUSION

The cultivation of coffee in Mexico is a line of economic, social and ecological relevance, however, the production of coffee in the country reports a decreasing trend. During the recent decade it was reduced to an annual average rate of 4.2%, to reach its lowest level since 1980 (FIRA, 2016).

The problems currently facing the Mexican countryside is not only coffee rust, but most of the coffee is grown in wooded areas and steep terrain, with topographies above 1,140 meters above sea level, making it very difficult to implement the technology used in other countries with flat lands, as well as the system of governmental support programs of several decades ago to boost the productivity of the favorable cultivation, the use of specialized varieties (Hausermann, 2014, 390-391), with monocultures in open sky or low shade that require application of Inorganic fertilizers and chemical control of pests, which triggers the impoverishment of the soil due to erosion, the leaching of nutrients and the rupture of the dynamics of natural self-regulation of pests, favored the production of plants in "nurseries" without regulated varieties by region, the traceability of the seed or the health of the growing medium, which has caused health problems such as the dispersion of root nematodes between the different territories, causing some producers to continue to perform grafts on robusta coffee patterns (which is resistant to nematodes), which although it increases production, it sometimes produces genetic variability through the appearance of chimeras and mutations, promoting a proliferation of varietal disorder (Contreras-Hernández, 2010, 146-148).

The production of coffee is strongly subject to supply-demand cycles of the product at the international level, being an evil that affects not only Mexico, but all other producing countries and has been presented throughout the history of the coffee trade. That is why the goal of all organizations and institutions linked to coffee is still high production, without seeing that it is not the only element of development, there are issues such as grain quality, productive integration, added value and competitiveness, which are ingredients for the success of any productive and commercial process, which has been relegated in its priority on its the effort to produce more coffee and at a lower cost, without considering that there are other options for the coffee crop to be profitable, Also, let's not forget that among other functions, coffee plantations stand out for their value in the conservation of biological diversity, their moderating role of drainage and infiltration of rainwater, storage of carbon dioxide and conservation and improvement of soil fertility; making it clear that conserving this type of agroecosystem is a priority (INIFAP, 2018).

So a proposal is to develop agroecological practices that allow the sustainable development of agriculture, the advancement towards inclusive and efficient food systems, promoting the protection of natural resources which meets the needs of the majority of producers, thus generating a virtuous circle between the production of healthy foods and the protection of natural resources (FAO, 2016, Contreras-Hernández, 2010, 149-158), providing to the coffee producing families in Mexico, especially small coffee producers, with best practices for the production of sustainable coffee, as well as publicize standards required by certifiers (organic, fair trade, sustainable agriculture, friendly to birds) so that they know, observe and can market their coffee as sustainable and organic coffee which its overall consumption is in greatest demand every day and has a higher price in the international market (Aranda-Bezaury et al., 2015, 21; Gutiérrez-Pérez et al., 2012, 100-105), using coffee residues to add value to the coffee crop either as fertilizer, biofuel, substrate for edible fungi or other products for various purposes, valued in the coffee industry, including the extraction of caffeine and polyphenols (Peña-Martínez et al., 2016).

In this way, not only can coffee be promoted, but it can also protect cloud forests that are in danger of disappearing (Williams-Linera et al., 2015, 96-98), with coffee plantations under the shade the conservation of regional biodiversity strategies, because they are shelters for it as well as intermediate points that facilitate the flow of individuals or genetic flow.

The main coffee regions of the state of Veracruz are rugged and wooded areas, where it is necessary to continue with traditional farming that requires more labor, at the same time, it has been shown that they promote the conservation of cloud forests, being necessary to give an agroecological approach to continue protecting this type of crop in the region, encouraging producers with varieties that are chosen not only for their production capacity but also according to the agroclimatic region, the resistance to pests and diseases, avoiding this way the use of unnecessary grafts in the area, as well as providing training to add value to coffee through the production of organic coffee, which is sustainable to the environment and the use of coffee residues. It will benefit the producer and the environment as well.

When analyzing the increase of consumers who are willing to pay more for gourmet coffee, another strategy that is proposed to increase the growth of the domestic market, are programs targeted at stimulating the "coffee culture" by increasing the consumption of the infusion of coffee from grains of first quality, grains with denomination of origin that assure their superior and outstanding quality, new flavors, new mixtures and environments that stimulate the overall consumption of this unique and exquisite product.

REFERENCES

Aguilera, Virginia, López-Espinoza, Antonio, Martínez, Alma Gabriela, Díaz, Felipe, Valdez, Elia, Galindo, Alma and De la Torre-Ibarra, Carolina. 2009. Effects of coffee consumption on feeding behavior in rats. *Diversitas: Perspectivas en Psicología,* 5:177-185. Accessed March 28, 2018. [*Diversitas: Perspectives in Psychology*]

Amaro-Rosales, Marcela and Gortari-Rabiela, Rebeca. 2016. Inclusive innovation in the Mexican agricultural sector coffee growers in Veracruz. *Economía Informa* 400:86-104. Accessed February 16, 2018.

AMECAFE. 2017. *Mexican Association of the Coffee Production Chain,* A. C. Accessed March 27 https://amecafe.org.mx/.

ANACAFE. 2014. *National Coffee Association.* Accessed March 27. https://www.anacafe.org.

Anunputtikul, Wantanee, and Rodtong, Sureelak (2004). Laboratory scale experiments for biogas production from cassava tubers. Paper presented at the *Joint International Conference on Sustainable Energy and Environment (SEE),* Hua Hin, Thailand, December 1–3.

Aranda-Bezaury, Josefina G., González-Bustamante, Benigno, Reyes-Santiago, Taurino. 2015. *Good practice guide for sustainable coffee,*

21-23. USAID, TNC, Rain Forest Alliance, The Goods Hole Research Center, Espacios Naturales, Desarrollo Sustentable.

ASERCA. 2018. *Agency of Services to the Marketing and Development of Agricultural Markets*. Accessed April 20. http://www.aserca.gob.mx/

Barreda-Abascal, Roberto, Molina, Luis, Haro-Valencia, Reyes, Alford, Chris and Verster, Joris C. 2012. Update on the effects of caffeine and its safety profile in food and beverages. *Revista Médica del Hospital General de México*, 75:60-67. Accessed February 25, 2018. [*Medical Journal of the General Hospital of Mexico*]

Belitz, Hans-Dieter, Werner Grosch and Peter Schieberte. 2004. *Food Chemistry*; 939-969. Springer: Berlin, Germany.

Benítez-García, Erika, José L. Jaramillo-Villanueva, Sergio Escobedo-Garrido, and Saturnino Mora-Flores. 2015. Characterization of coffee production and trade in the municipality of Cuetzalan, Puebla. *Agricultura, Sociedad y Desarrollo* 12:181-198. Accessed April 23, 2018. [*Agriculture, Society and Development*]

Bermúdez Savón, Catalina, Rosa, García Oduardo, Nora, Agbozouhoue, Kodjovi Kekeli and Alberni, Migdalia Serrano. 2018. Productivity evaluation of two strains of Pleurotus on coffee pulp *coffea canephora pierre ex frhoener. Tecnología Química*, *38*:246-255. Accessed May 30, 2018. [*Chemical Technology*]

Blandón-Navarro, Lorena, Sandra, Castillo, Bayardo and Llancys López, Ana 2013. Validation of coffee mucilage for the production of ethanol and organic compost. *El Higo Revista Científica*, 3:10-13. Accessed April 29, 2018. [*The Higo Scientific Review*]

Canet-Brenes, Guillermo, Soto-Víquez, Carlos, Ocampo-Thomason, Patricia, Rivera-Ramírez, Javier, Navarro-Hurtado, Alejandra, Guatemala-Morales, Guadalupe M. and Socorro Villanueva-Rodríguez, S. 2016. *Status and trends of coffee production in Latin America and the Caribbean*. 148-150. Instituto Interamericano de Cooperación para la Agricultura (IICA).

Castellar-Ortega, Cecilia, Grey, Cely-Bautista, María Mercedes, Cardozo-Arrieta, Beatriz María and Rosales-Díaz, Cristian C. L. 2017. Use of the coffee waste for the production of activated carbon and its evaluation in the adsorption of the direct marine blue dye. Paper presented at the *Memorias III Seminario Internacional de Ciencias Ambientales SUE-Caribe*. Barranquilla, Colombia june 28-29.

CEDRSSA. 2016. *Center of Studies for Sustainable Rural Development and Food Sovereignty*. Accessed April 19, 2018. http://www.cedrssa.gob.mx.

CENICAFE. 2018. *National Coffee Research Center*. Accessed May 21 https://www.cenicafe.org.

Central América Data. 2018. Accessed April 21. https://www.centralamericadata.com.

Central América Data. 2018. Accessed May 21. https://www.centralamericadata.com.

Certimex. 2018. *Mexican Certifier of Ecological Products and Processes, S.C.* Accessed April 23. https://www.certimexsc.com.

Challenger, A. 1998. *Use and conservation of terrestrial ecosystems of Mexico: past, present and future*, 847-848. CONABIO-Instituto de Biología UNAM-Sierra Madre. México, D. F.

CONAPO. 2015. *National Population Council* Accessed April 19, 2018. http://www.gob.mx/conapo.

CONEVAL. 2018. *National Council for the Evaluation of Social Development Policy* Accessed April 21, 2018. https://coneval.org.m.

Contreras, Armando, y Gerardo Hernández-Martínez. 2008. *Smells good, better know: the producer organization project Biocafé*. 9-10. Instituto de Ecología, AC.

Contreras-Hernández, Armando. 2010. Coffee plantations of Veracruz and its contribution to sustainability. *Estudios Agrarios*, *45*:143-161. Accessed March 19, 2018. [*Real Estate Agencies*]

Coutiño-Puchulia, Victor, Santoyo-Cortésa, Vinicio H., Flores Verduzcoa, Juan José and Muñoz-Rodríguez, Manrrubio 2017. Comparative analysis of two organizations of small coffee producers in Oaxaca, Mexico. *Rev. Transitare*. 3:41-57.

DaMatta, F. M., Ronchi, C. P., Maestri, M. & Barros, R. S. 2007. Ecophysiology of coffee growth and production. *Brazilian journal of plant physiology*, *19*(4), 485-510.

Declaratoria General de Protección de la Denominación de Origen "*Café Chiapas*". 2003. Accessed June 09, 2018. http://www.wipo.int/wipolex/es/text.jsp?file_id=220905. [General Declaration of Protection of the Denomination of Origin "Café Chiapas".]

Declaratoria General de Protección de la Denominación de Origen "*Café Veracruz*." 2000. Accessed June 09, 2018. http://www.wipo.int/edocs/lexdocs/laws/es/mx/mx078es.pdf. [General Declaration of Protection of the Denomination of Origin "Café Veracruz."]

Díaz, P, y Riaño, C. E. 2017. Characterization of roasted coffees, from the E-220A ecotope. *Nova, 15:*33-43. Accessed March 25, 2018.

Dicovskiy-Rioboo, Luis María, Pichardo-Hernández, Claudio Benito, Rodríguez-Zeledón, Aleyda Nohemí, Johexi Martínez Centeno, Bayardo Johexi and del Rosario Rodríguez Espinoza, Kiara. 2014. Validation of briquettes made with coffee husks for the replacement of firewood in the production of corn donuts. *Rev. El Higo*. 4:10-19. Accessed May 18, 2018.

Duangjai, Acharaporn, Suphrom, Nungruthai, Wungrath, Jukkrit, Ontawong, Atcharaporn, Nuengchamnong, Nietra and Yosboonruang, Atchariya. 2016. Comparison of antioxidant, antimicrobial activities and chemical profiles of three coffee (*Coffea arabica* L.) pulp aqueous extracts. *Integr. Med. Res*. 5: 324-331. Accessed May 10, 2018.

Echeverri, Dario, Buitrago, Lorena, Montes, Félix, Mejía, Ingrid and del Pilar González, María. 2005. Coffee for cardiologists. *Rev. Colomb. Cardiol.* 11:357-365. Accessed February 28, 2010.

Escamilla-Prado, Esteban, Ruiz-Rosado, Octavio, Zamarripa-Colmenero, Alfredo and González-Hernández, Víctor A. (2015). Quality varieties of organic coffee in three regions of Mexico. *Revista de Geografía Agrícola,* (55), 45-55. Accessed March 18, 2018.

Espinosa-García, Antonio, José, Uresti-Gil, Jesús, Vélez-Izquierdo, Alejandra, Moctezuma-López, Georgel, Uresti-Durán, Diana, Góngora-González, Sergio Fernando and Inurreta-Aguirre, Héctor Daniel. 2016. Productivity and potential profitability of coffee (*Coffea arabica* L.) in the Mexican tropics. *Revista mexicana de ciencias agrícolas*, 7(8), 2011-2024 Accessed April 28, 2018.

Euromonitor International. 2017. Accessed May 21, 2018. http://www.euro monitor.com/.

FAOSTAT. 2017. Accessed June 29, 2018. http://www.fao.org/faostat/en/#data/QC.

FAO. 2016. Accessed June 29, 2018. http://www.fao.org.

Figueroa-Hernández, Esther, Pérez-Soto, Francisco and Godínez-Montoya, Lucila 2016. *The coffee market in Mexico*, 68-71. ©ECORFAN, Texcoco de Mora-México.

FIRA. 2016. Accessed March 07, 2018. https://www.fira.gob.mx.

Fonseca-García Libia, Calderón-Jaimes, Lilia S. and Rivera, María E. 2014. *Antioxidant capacity and total phenolic content of coffee and coffee by-products produced and marketed in Norte de Santander (Colombia)*. Vitae 21:3, 228–236. Accessed April 18, 2018.

FUNDESYRAM. 2010. Accessed April 13, 2018. http://www.fundesyram.info.

Giraldo-Quintero, Juan José, Carlos David Niño-Méndez, and Zulma Vianchá-Sánchez. 2017. Analysis of best practices in the process of coffee processing: study experience in the municipality of Viotá (Cundinamarca, Colombia). *Ingeniería Solidaria*. 13(22): 121-135. Accessed May 28, 2018.

González de Mejía, Elvira, and Ramírez-Mares, Marco Vinicio M. V. 2014. Impact of caffeine and coffee on our health. *Trends in Endocrinology & Metabolism*, *25*(10), 489-492. Accessed May 11, 2018.

Gotteland, Martín, and Saturnino de Pablo. 2007. Some truths about coffee. *Revista chilena de nutrición*, 34:105-115. Accessed June 02, 2018. [*Chilean nutrition magazine*]

Gutiérrez-Pérez, Cynthia, Tuñón-Pablos, Esperanza, Limón-Aguirre, Fernando, Morales, Helda and Nielsen, Ronald Nigh. 2012. Social representations of organic food among consumers in Chiapas. *Estudios sociales (Hermosillo, Son.)*, *20*(39), 99-130. Accessed June 18, 2018.

Haile, Mebrathu. 2014. Integrated volarization of spent coffee grounds to biofuels. *Biofuel Research Journal*, 1:65-69. Accessed May 29, 2018.

Hauserman, Heidi. 2014. Maintaining the Coffee Canopy: Understanding Change and Continuity in Central Veracruz. *Hum. Ecol.,* 42: 381-394. Accessed June 13, 2018.

Hernández-Cázares, A. S., Real-Luna, N., Delgado-Blancas, M. I., Bautista-Hernández, L. and Velasco-Velasco, J. 2016. Agro-industrial waste with potential for composting. *Agroproductividad*, 9(8): 10-17. Accessed May 15, 2018.

Hurtado, Ifigenia, and Herrera-Pinilla, Juan Carlos 2013. Estudio de la diversidad morfológica y molecular del germoplasma de Coffea liberica bull ex hiern en Colombia. *Revista Cenicafé* 64 (2): 31-47. Accessed March 30, 2018. [Study of morphological and molecular diversity of the germplasm of *Coffea liberica* bull ex hiern in Colombia. *Cenicafé Magazine*]

Hurtado-Rengifo, Pablo José. 2016. Use of secondary biomass waste as fuel. *Revista Ingenium*, 1:26-30. Accessed May 12, 2018. [*Ingenium Magazine*]

ICAFE. 2014. Accessed April 29, 2018. http://www.icafe.cr.

ICO. 2017 *International Coffee Organization. Annual Review 2016/2017*, Accessed May 21, 2018. http://www.ico.org.

ILO Foundation. 2017. *International Labour Organization*. Accessed March 09, 2018. http://www.ilo.org.

IMPI. 2018. Accessed April 17. https://www.gob.mx/impi.

INFOAGRO. 2017. Accessed May 25, 2018. http://www.infoagro.com/.

Inforural, 2018. Accessed May 21, 2018. https://www.inforural.com.mx/.

INIFAP. 2018. Accessed May 05. https://www.inifap.gob.mx.

Intercontinental Exchange (ICE). 2018. Accessed May 13 https://www.intercontinentalexchange.com.

Janissen, Brendan, and Huynh, Tien. 2018. Chemical composition and value-adding applications of coffee industry byproducts: A review. *Resources, Conservation and Recycling,* 128:110–117. Accessed May 04, 2018.

Karmee, Sanjib Kumar. 2017. A spent coffee grounds based biorefinery for the production of biofuels, biopolymers, antioxidants and biocomposites. *Waste Management*, 128:240-254. Accessed June 11, 2018.

Kim, Jaai, Kim, Hakchan, Baek, Gahyun and Lee, Changsoo. 2016. Anaerobic co-digestion of spent coffee grounds with different waste feedstocks for biogas production. *Waste Management*, 60:322-328. Accessed May 28, 2018.

Koby, Ross. 2015. Chemical Valorization of Coffee Bean Waste. *McGill Green Chemistry Journal*, 1:95-98. Accessed June 23, 2018.

Kovalcik, A., Kucera, D., Matouskova, P., Pernicova, I., Obruca, S., Kalina, M., Enev, Vojtěch, and Marova, I. 2018. Influence of removal of microbial inhibitors on PHA production from Spent Coffee Grounds employing Halomonas halophila. *Journal of Environmental Chemical Engineering*, 6:3495–3501. Accessed May 28, 2010.

Lede, Silvia. (2008). *Los Biocombustibles*. Accessed June 17, 2018. http://www.argenbio.org/. [*Biofuels*]

León, J. 2000. *Botany of tropical crops*. 350-364. IICA. San José, CR.

López-García, Francisco Javier, Escamilla-Prado, Esteban, Zamarripa-Colmenero, Alfredo and Cruz-Castillo, J. Guillermo. 2016. Production and quality varieties of coffee (*Coffea arabica* L.) in Veracruz, Mexico. *Rev. Fitotec. Mex*, 39:297–304. Accessed May 24, 2018.

Manals-Cutiño, Enma M., Salas-Tort, Dolores and Penedo-Medina, Margarita. 2018. Characterization of plant biomass "coffee husk." *Tecnología Química*, *38*(1), 169-181. Accessed June 14, 2018. [*Chemical technology*]

Marín-García, Carolina, and Puerta-Quintero, Gloria Ines. 2008. Chlorogenic acids content in grains *Coffea arabica* and *C. canephora* L. as fruit development. *Cenicafe*, 59:7-28. Accessed April 14, 2018.

Medina-Meléndez, Alfredo, José, Ruiz-Nájera, Ramiro Eleazar, Gómez-Castañeda, Julio César, Sánchez-Yáñez, Juan Manuel, Gómez-Alfaro, Gabriela and Pinto-Molina, Omar. 2016. Study of the coffee production system (*Coffea arabica* L.) in the Frailesca region, Chiapas. *Ciencia UAT*, *10:*33-43. Accessed May 25, 2018.

Mehari, Bewketu, Redi-Abshiro, Mesfin, Chandravanshi, Bhagwan Singh, Combrinck Sandra and McCrindle, Rob. 2016. Characterization of the

Cultivation Region of Ethiopian Coffee by Elemental Analysis. *Analytical Letters,* 18:1-.

MIDA. 2015. Accessed March 03, 2018. https://www.mida.gob.pa/upload/documentos/ficha_de_cafe_2015.pdf.

Montagnini, Florencia, Somarriba, Eduardo, Murgueitio, Enrique, Fassola, Hugo and Eibl, Beatriz. 2015. *Agroforestry systems productive, socioeconomic and environmental functions,* 5-20. Serie técnica. Informe técnico/CATIE. Colombia.

Montero-Alpírez, Gisela, Jaramillo-Colorado, Beatriz E., Vázquez-Espinoza, Ana M., Coronado-Ortega, Marcos A., Gacía-González, Conrado and Toscano-Palomar, Lydia. 2017. Experiences of waste use for the generation of biodiesel in Colombia and Mexico. *Revista Internacional de Contaminación Ambiental*, 32:77-90. Accessed June 14, 2018. [*International Journal of Environmental Pollution*]

Montilla-Pérez, Jimena, Arcila-Pulgarín, Jaime, Aristizábal-Loaiza, Manuel, Montoya-Restrepo, Esther C., Puerta-Quintero, Gloria I., Oliveros-Tascón, Carlos E. and Cadena-Gómez, Gabriel. Characterization of some physical and conversion factors properties of coffee during the traditional benefit. *Cenicafé*, 59:120-142. Accessed June 03, 2018.

Montoya-Ferrer, Jaime, and Montoya-Vázquez, Vanessa. 2011. Coffee special coffee growing alternative for Risaralda. *Revista Académica e Institucional*, 89:93-105. Accessed May 28, 2018. [*Academic and Institutional Magazine*]

Muñoz-Muñoz, Deyanira, Pantoja-Matta, Alvaro and Cuatin-Guarin, Milton Fernando. (2014). Use of agro-industrial waste as biofuel and biorefinery. *Biotecnología en el Sector Agropecuario y Agroindustrial*, 12:10-19. Accessed June 19, 2018. [*Biotechnology in the Agricultural and Agroindustrial Sector*]

Mussatto, Solange I., Machado E., Ercília M. S., Martins, Silvia and Teixeira José A. J., 2011. Production, composition and application of coffee and its industrial residues. *Food and Bioprocess Technol.*, 4:661-672. Accessed June 15, 2018.

Mussatto, Solange I., Ballesteros, Lina F., Martins, Silvia, Teixeira, José A. 2012. Extraction of antioxidant phenolic compounds from spent coffee grounds. *Separation and Purification Technology*. 83, 173-179. Accessed June 19, 2018.

National Agricultural Planning of Mexican Coffee. 2018. Accessed May 11. https://www.gob.mx/cms/uploads/attachment/file/256426/B_sico-Caf_pdf.

National Federation of Coffee Growers of Colombia. 2018. Accessed March 30. http://www.federaciondecafeteros.org.

Noguera, R. R. y Posada, S. L. (2015). Effect of drying method on digestibility in situ pulp coffee (*Coffea arabica*). *Livestock Research for Rural Development,* 29:1-7. Accessed June 09, 2018.

Noriega-Salazar, Adrianyela, Silva-Acuña, Ramón and de Salcedo, Moraima García. 2008. Use of coffee pulp in animal feed. *Zootecnia Tropical*, 26:411-419. Accessed May 28, 2018.

Office of Energy Efficiency and Renewable Energy. 2018. Accessed June 10. https://www.energy.gov/eere/bioenergy/waste-energy.

Oliveira, Marta, Ramos, Sandra, Delerue-Matos Cristina and Morais, Simone. 2015. Espresso beverages of pure origin coffee: Mineral characterization, contribution for mineral intake and geographical discrimination. *Food Chemistry*. 177: 330-338. Accessed April 02, 2018.

Ortega-Hernández, Alejandro, and Ramírez-Valverde, Benito. 2013. Crisis of coffee growing and migration in the context of poverty and marginalization. The case of the indigenous producers of Huehuetla, Puebla. *Ra Ximhai*, 9:173-186. Accessed May 17, 2018.

Pardo-Lozano, Ricardo, Álvarez-García, Yolanda, Barral-Tafalla, Diego, and Farro-Albaladejo, Magí. 2007. Caffeine: a nutrient, a drug, or a drug of abuse. *Revista Adicciones*, 19:225-238. Accessed April 21, 2018. [*Addictions Magazine*]

Peña-Martínez, Andrea, Paula, Ramírez-Saldarriaga, Natalia Andrea and Rúa, María Yuliana. 2016. Carbonated beverage processing based pulp and cisco Arabica coffee (*Coffea arabica* L.). *Agronomía Colombiana,* 34:S1175-S1177, Accessed May 18, 2018.

Pérez-Akaki, Pablo, and Pérez-Tapia, Manuel. 2011. The denominations of origin of Mexican coffee and its questions as a model of regional development. *Geocalli*, 21:1-61. Accessed May 28, 2018.

PROFECO. 2001. Accessed June 11, 2018. https://www.profeco.gob.mx/revista/pdf/est_01/Cafe.pdf.

PROY-NMX-F-173-1999. Accessed June 13, 2018. https://www.colpos.mx/bancodenormas/nmexicanas/NMX-F-173-1999.PDF.

Puerta-Quintero, Gloria Ines, and Ríos-Arias, Sara. 2011. Chemical composition of the coffee mucilage, according to the time of fermentation and refrigeration. *Cenicafé*, 62:23-40. Accessed April 25, 2018.

Quintero-Castaño, Víctor, Cárdenas-Valencia, Carlos and Lucas-Aguirre, Juan. 2014. Characterization of plant waste generated in the wholesale distribution center of the city of Armenia (MERCAR) for industrial use in the production of biofuels. *Scientia et Technica Año XIX*, 19:261-267. Accessed June 16, 2018.

Ramalakshmi, K., Rao, L. J. M., Takano-Ishikawa, Y. and Goto, M. 2009. Bioactivities of low-grade Green coffee and spent coffee in different in vitro model systems. *Food Chemistry* 115, 79-85. Accessed June 18, 2018.

Rathinavelu, Rajkumar, and Graziosi, Giorgio. 2005. Possible alternative uses of coffee wastes and by-products. *Organización Internacional del Café*. 5:1-4. Accessed June 28, 2018. [*International Coffee Organization*.]

Rodríguez-Valencia, Nelson. 2013. Alcohol production from coffee pulp. *Cenicafé*, 64(2):78-93. Accessed May 27, 2018.

Rodríguez-Valencia, Nelson., Zambrano-Franco Diego Antonio and Ramírez-Goméz, Cesar Augusto. 2013. "Management and disposal of by-products and wastewater from the coffee benefit." In *Manual del cafetero colombiano: Investigación y tecnología para la sostenibilidad de la caficultura tomo III, [Manual of the Colombian Coffee Grower: Research and Technology for the Sustainability of Coffee Production, Volume III]* edited by Gast, Fernando, Pablo Benavides, Juan Rodrigo Sanz, Juan Carlos Herrera, Víctor Hugo Ramírez, Marco A.

Cristancho, and Sandra Milena Marín, 111-136 Chinchiná, Colombia. Accessed May 25, 2018.

Romero-Arenas, Omar, Martínez-Guerrero, Marco A., Damián-Huato, Miguel, Ramírez-Valverde, A. Benito and López-Olguín, Francisco. 2015. Shiitake mushroom production (shiitake Pegler) using synthetic blocks agroforestry residues. *Revista mexicana de ciencias agrícolas*, *6*:1229-1238. Accessed May 16, 2018. [*Mexican Journal of Agricultural Sciences*]

Ruelas-Monjardín, Laura C., Nava-Tablada, Martha Elena, Cervantes, Juan, and Barradas, Víctor L. 2014. Environmental importance of coffee agroecosystems under shade in the central mountainous area of the state of Veracruz, Mexico. *Madera y bosques*, *20*(3), 27-40. Accessed April 20, 2018. [*Wood and forests*]

SAGARPA. 2017. Accessed April 03, 2018. https://www.gob.mx/sagarpa.

Saucedo-García, Aurora, Anaya, Ana Luisa, Espinosa-García, Francisco J., and González, María C. 2014. Diversity and communities of foliar endophytic fungi from different agroecosystems of *Coffea arabica* L. in two regions of Veracruz, Mexico. *PloS One*, *9*(6), e98454. Accessed February 28, 2010.

Schüßler, Lennart. 2009. Protecting "single-origin coffee" within the globla coffee market: The role of geographical indications and trademarks. *The Estey Centre Journal of International Law and Trade Policy*. 10(1): 149-185. Accessed May 14, 2018.

Sepúlveda, Wilmer. S., Chekmam, Louiza, Maza, María T. and Mancilla, Nelson O. 2016. Consumers' preference for the origin and quality attributes associated with production of specialty coffees: Results from a cross-cultural study. *Food Research International*, *89*, 997-1013. Accessed June 01, 2018.

SIAP. 2017. Accessed April 19, 2018. https://www.gob.mx/siap.

Sosa-Fernández, V., López-Morgado, R., Toledo-Aceves, T. and Bárcenas-Pazos, G. 2017. Opportunities for conservation of the cloud forest through alternative management: coffee agrosystems. *Agro-productividad, 10:62-67*. Accessed April 12, 2018.

Spanish Coffee Federation. 2018. Accessed March 09, 2018. http://www.federacioncafe.com.

Taramona-Ruiz, Luis Alberto, Barrena-Gurbillón, Miguel Angel, Gamarra-Torres, Oscar Andrés, Choy-Wong, Maximiliano and Rodríguez-Ponce, Yoandro. 2017. Biogas yield and socioeconomic benefits in sustainable rural development of the Andean communities. Ancash-Perú. *Revista de Investigaciones de la Universidad Le Cordon Bleu*, *4*(2), 49-57. Accessed April 30, 2018. [*Research Magazine of the Le Cordon Bleu University*]

Tian, Rong-Rong, Pan, Qiu-Hong, Zhan, Ji-Cheng, Li, Jing-Ming, Wan, Si-Bao, Zhang, Qing-Hua and Huang, Wei-Dong. 2009. Comparison of phenolic acids and flavan-3-ols during wine fermentation of grapes with different harvest times. *Molecules*. *14:*827-838. Accessed May 04, 2018.

USDA. 2017. *U.S. Department of Agriculture*. Accessed April 11, 2018. https://www.usda.gov.

Valencia-Sandoval, Karina, and del Rosario García-Velázquez, María. 2017. The process of coffee production in the region of the Gulf of Mexico. In *Boletín Científico de las Ciencias Económico Administrativas del ICEA*, edited by Duana-Ávila, Danae, https://doi.org/10.29057/icea.v6i11.2707. [*Scientific Bulletin of the Economic and Administrative Sciences of the ICEA*]

Valencia-Sandoval, Karina. 2015. Competitiveness of coffee exports from Colombia, Guatemala and Mexico to the US market (2001-2014). *CIENCIA ergo-sum*, 23:239-246. Accessed May 25, 2018. Accessed April 06, 2018.

Vardon, Derek R., Moser, Bryan R., Zheng, Wei, Witkin, Katie, Evangelista, Roque L., Strathmann, Timothy J., Rajagopalan, Kishore. and Sharma, Brajendra K. 2013. Complete Utilization of Spent Coffee Grounds To Produce Biodiesel, Bio-Oil, and Biochar. *ACS Sustainable Chem. Eng*. 1:1286–1294. Accessed May 22, 2018.

Vidal, Fernando, Pastor, Jesús T., Aparicio, Juan and Ortiz, Lidia. 2017. Technical efficiency of protected designations of origin in Spain: An analysis by type of wine marketed. Paper presented at the *40th World Congress of Vine and Wine*, Sofia, Bulgaria, 29 May - 02 June.

Waller J. M., Bigger, M. and Hillocks, R. 2007. *Coffee Pests, Diseases and Their Management, 325–335*, CABI, Wallingford, UK.

Williams-Linera, Guadalupe, López-Barrera, Fabiola and Bonilla-Moheno, Martha 2015. Establishing the baseline for the restoration of the cloud forest in a peri-urban landscape. *Madera y Bosques*, 2:89-101. Accessed February 28, 2010. [*Wood and Forests*]

Zuorro, Antonio, and Lavecchia, Roberto. 2012. Spent coffee grounds as a valuable source of phenolic compounds and bioenergy. *Journal of Cleaner Production*. 34, 49-56. Accessed June 19, 2018.

INDEX

C

D

E

F

G

H

L

M

N

O

P

R

S

T

U

V

W